BELL-SHAPE
TESTING SYSTEM

Bell-Shape Testing System

Testing the Students Based on
Simple and Complex Teachings
Related to Bloom's Taxonomy
of Educational Objectives

Acene Fleurmons, BSW, MOM, and EdD

To order additional copies of this book, contact:
Xlibris
1-888-795-4274
www.Xlibris.com
Orders@Xlibris.com
715393

CONTENTS

PREFACE

This book is about a presentation of Benjamin Bloom's "*Taxonomy of Educational Objectives Book 1 Cognitive Domain*". It rather wants to be a research paper in which I make a profound reflection on the educational objectives presented by Bloom in 1956. I take the opportunity to seek knowledge or information on how they are implemented by the schools. The greatest opportunity I have is to indicate how these educational objectives should be implemented in lifelong learning so the students of any age, especially in the public schools, can have insight of them for their success.

This book also contains some critics of Bloom's text related to the classification of the objectives. For example, comprehension cannot be classified immediately after knowledge because one needs to develop some mental and intellectual efforts before he or she can be confident with having insight into anything. This stage of knowing is based on the analysis of the encountering fact.

In fact, this book is about a proposal, a new and innovative method of assessing the students in a fair manner, which will fill up an educational gap that one can view only if he or she has a vision of students' success. However, it is impossible to lead them to that state if they are tested unfairly, not taught properly, and not taught based on the state-given curriculum. I intend to look for equilibrium, a balance between teaching and its content and the testing system. Out of this balance, the student will always feel the exams' pressure. This is why I came up with the bellshaped testing system to help kids breathe, being free from the pressure of exams, so they can, at least and at last, show they are able to improve as Benjamin Bloom wished it.

That system of testing is called "bell-shaped testing system." Improvement may be really effective when teachers and state representatives properly use that testing system offered in this research. It states why students should be tested this way indicated in it, including using content based on the pre-established curriculum, a psychological aspect, and how to shape the structure of the test in itself, which establishes a clear difference between simple and complex notions or memory and skill questions. This method of testing kids should not be isolated; it should be preceded by a method of teaching covering not only the curriculum but also the integrated participation of students in class presentations. At this level, a recommendation is given as a salutary advice: kids or students at all levels should not be tested based on concepts that have not been taught.

This book highlights a lot of educational theories and teaching methods. It is an occasion to emphasize Bloom's instructional method, which is *mastery learning*, and that of Hunter, called *mastery teaching*. These educational theorists contended that all kids have the ability to learn at the highest level possible. In order for this to happen, they must be taught properly using pedagogical technology that can make them understand, using formative assessment techniques and providing them with feedback.

In the same perspective of teaching kids based on certain workable instructional methods and strategies that can lift them up to fully understand the content of each session and help provide a complete educational package to them, I present in this book a teaching strategy containing nine instructional principles:

1. Choosing a teaching method
2. Teaching with the educational goal in your mind in general and the goal of each course and detailed objectives thoroughly present
3. Teaching based on the designated curriculum
4. Teaching to help students with storing and recall
5. Teaching for comprehension or understanding
6. Using regular feedback to check for understanding
7. Assessing the students based on the teaching content
8. Using cooperative learning
9. Regularly assessing the students using a formative assessment method (chapter VI).

After an explanation of these instructional principles, an empirical research has been conducted with the purpose of checking out the kind of method that teachers use in their daily practice of teaching and how kids are effectively doing. My hypothesis is the following:

If the teachers teach based on the established curriculum and use a good instructional method, students who are tested only based on what they have learned from their teachers and also based on the bell-shaped testing system—where questions are arranged from the bottom to the top, from simple to complex questions, or from retrieval to synthesis questions—will always earn grades no lower than 80.

It was found that teachers are not using a standardized teaching method and they have no teaching strategy to make their students understand them very well. Lastly, they do not assess the kids based on the teaching content, and they are slipping away from the prescribed curriculum. Moreover, teachers and state assessors do not structure the assessment content as appealing to a happy and relaxed state of mind, favoring the kids. Instead, they give them tests with the objective of destroying their combative spirit and resilient mind to courageously face tough exams within the intended program they have been taught. Consequently the students repeatedly fail to be successful in school.

This failure is interpreted as not the students' failure, but that of the teachers and the system. It is like they've been unconsciously assessing themselves. Fortunately, we still can remedy this threatening and frightening situation by implementing the bell-shaped testing system and incorporating advice given in the end of the research as recommendations. We, Educational Services and Research Center, are pretty sure that it is not too late.

Acene Fleurmons, BSW, MOM, and EdD
Founder and President of Educational
Services and Research Center
Fleurmons307@yahoo.com

INTRODUCTION

Education is a concept that has a variety of meanings to different people from different cultures, related to their conception of the world, life, and the development of human beings, which includes mind, body, and spirit. However, regardless of the people's diverse worldviews of education, it remains an eminent, prominent, and a basic element for human development. It is based on it that someone learns to speak and to exchange with others in his or her environment; it is based on it that that same person learns how to satisfy his or her primary and extended needs, whether they are psychological, security, love, and self-actualization (Abraham Maslow, 1968).

It is also based on education that human beings learn how to coexist and share a common space in order to create a family, a neighborhood, a community, a city, a county, a state, and a country. At any level, they altogether form what we call a society. As the animals are also able to create one that could be as organized as ours (Ronald Heifetz, 1994), it is extremely important that we educate ourselves and use our reason just to make a difference. Therefore we cannot deny the value of education in our lives and our society, civilized or not. This is why writers, poets, sociologists, anthropologists, social workers, educators, and philosophers cannot stop thinking about it and debating on it as it is our greatest concern in this research paper.

As it is said, without education our lives would have no sense and meaning at all, and our actions and organizations would be chaotic. Unfortunately, some societies are still not organized, and their educational systems are weak because they are not able to put into play substantial instructional elements and necessary educational leadership to shift their societies from chaos to a higher expected level, such as Haiti and Somalia. However, some countries—such as the USA, Canada, and those in Europe—already have sophisticated educational systems for the advancement of their people, their economy, their health system, and their political activities are fruity. It is true that the development of those parameters is totally dependent upon the level of a country's educational system.

As we recognize the value of education in a country, one can easily determine the level of its development based on its educational system. The educational system of that country may be informal, formal, or both. It is informal if its people learn to do things out of an established instructional system, such as public schools, charter schools, or private schools, as it is in the United States today. On the contrary, the educational system of that country is formal if it is institutionalized, and institutionalized education is constructed based on the country's culture (beliefs, arts, and social institutions), ethnicity, race (common ancestry and language), and the country's willingness to cut the influence of tradition in order to build a rationalized modern and/or postmodern society. All those are possible based on three elements. The first element is education, the second element is again education, and the third one is still education.

In the introductory part of the research "Bell-Shaped Testing," it is good to think about how education helps individual growth and how to acknowledge it as fact. This acknowledgement can lead us to the identification of educational goals and objectives, which should be set once in order to be a guide through this interesting educational research. Not only this, educational objectives can indicate what educational institutions do to make sure that students attain a certain expected level of achievement.

The School, Individual Growth, and Society's Development

The school is a formal institution where children learn to read and write properly. It is also where they learn to solve life's threatening problems. School is where immense intellectual activities take place, those that have to see with the needs of the students, the environment, the law, and with practical morale. It is a place where individuals learn to be good and become active citizens. In other words, they learn to be accountable and reliable while growing up and during their different developmental stages.

The Individual's Growth

There are two types of growth: physical and spiritual. Physical growth is related to the individual's body structure, the hormonal development that makes the individual become a woman and or a man. This development is called physical maturity, moving from infanthood to adulthood. During this time, children learn at home how to take care of themselves, including taking a shower, brushing their teeth, how to dress themselves, and how to cook, and eat quality and healthy foods—as educational experts recognize that eating habits affect academic performance. It is impossible for them to grow up properly without those primary lessons and healthy foods intake. The home environment has a very great influence on them; this is why their home needs to provide them with material elements that should allow them to satisfy their primary or physical needs so they can grow up as expected. Otherwise, they may encounter difficulties in their learning process at school, their second and spiritual home, where they can learn schemes of behaviors and how to control their emotions by rational means so they can have a balanced life based on the normal development of their body, mind, and spirit.

Spiritual growth can be called intellectual growth. It is a fact that kids learn to solve simple and complex problems based on their ages—from zero to eighteen years of age. At age zero, infants are introduced to the physical world, and they interpret it in their manner; they are considered as self-learners. Therefore, their mothers need to create highly rich and healthy environments for them so they can see, touch, smell, and taste. We should not forget that the learning process starts at this age and ends on the day of their death. As anticipation, we will reconsider the beginning of learning process at the age of fetal learning (3 to 6 month before birth) in our next research on learning and the affective approach as a good start to demonstrate that there is no intelligence differences. The learning process is a lifelong learning.

At the age of three, kids should be able to go to school to learn how to react to their environment safely and rationally by using their reason. The school should be able to create architectonic and balanced curricula to facilitating the flawless development of the kids' intellectual growth from stage to stage.

When we look at Erickson's psychosocial stages of development from infancy to adolescence and young adulthood, we can find a way to relate to different schooling ages necessary for the development of the individual, as Erickson did himself. Three stages may be at the center of our focus: preschooler (from ages three to five, corresponding to Initiative vs. Trust), elementary school (from ages six to eleven, period corresponding to Competency vs. Inferiority), and adolescence (teen stages, corresponding to Identity vs. Confusion). Individuals express different needs at each of these stages. Therefore scholarly curricula must be built to satisfy those needs.

The Fundamental Goals of Education

The ultimate goal of education is to help human beings have a distance from the state of nature by means of using their reason so they can live harmoniously in their environment. Human beings are called to dominate the nature. In order to do that, they should be educated to understanding themselves as human beings—the complexity of their body and their mind, what causes them to have somatic and psychological needs, and, at last, how to satisfy them. As they cannot do it alone, there is necessity for the existence of social and instructional institutions (John Dewey, 1916) to help them out. Those socio-instructional institutions (educational TV shows, social or social work organizations, counseling centers, and counseling oriented churches) are able to produce individuals who can build balanced societies and create the means for life renewal. Dewey (1916) thinks of education as a social function, a direction, something both conservative and progressive, and as growth.

Our kids need direction to learn about their ego and to recognize that they are not only the ones that occupy the space where they are, that some other people like them share that common space with them. Knowing their rights and their boundaries is essential. We actually listen to kids and young people who take pleasure in speaking about their freedom, allowing them to do whatever they want under no restriction. That is false; one can do whatever he or she wants within the law and acceptable mores and cultural principles. Teaching the notions of altruism, sharing, and living as a group is a necessity. It is also important to teach them about self, value, democracy, and education. It is imperative that they know why they are not home and the necessity of education and what one can gain from it. All these form what educational theorists, sociologists, anthropologists, and philosophers may call socialization from the early ages to adolescence. This is where the first fundamental goal of education lies. Failure to get an insight into this is failure in life, and the risk of living at the state of nature (Montesquieu, 1648, in his *The Spirit of the Laws*) could be an unavoidable reality, as Thomas Hobbes said (state of war) in his book called *Leviathan* (chapter XV), published in 1651 and republished by Barnes & Noble in 2004. We are sure that our society is over this, but we have to continuously sustain it by means of education.

The concept of socialization is associated with individual role play as all societies are primarily functional. Kids should feel that they belong somewhere and that they are integrated and playing their role so they can develop a spirit of belonging to help them proudly learn better and live based on certain social norms. They should also learn about the importance of playing an authentic role and playing it effectively so they become eager to play one or to perceive one. These teachings form the dorsal spine of personality and the earliest perception of identity. From there, kids would be able to identify themselves authentically and learn to put aside infantile behaviors to adopt some early youth's notion of maturity. If a kid cannot perceive the world as such at twelve, it is very possible for him or her to enter adolescence with some kind of identity crisis. Learning what is right, or not, moral or immoral, condemnable or not, permissible or not is uncontestable. To suffer and to stand up for what is right are top educational demands at the moment. At this point, Dewey (1991) perceived some kind of childhood curriculum, and we think that this is extremely crucial.

Kids Need a Great and Open Environment

According to Peter Russell in his book titled *"The Brain Book"*, kids need a great and rich environment in order to reach the potentials of their mind and their brain since their birthday. It is possible for a child who lives in an environment fostered by educated parents to accomplish greater things than one who grows up in a poor environment. Parents should start educating their children in their cribs. This is the moment to start the process of making a child a star, and the schooling institutions, later, would have the right to bring them to brighter light, to the finish line.

How can a kid at the age of four read two books per day, then skip some elementary classes and the entire high school to enter college? How can he become an advanced math teacher in college at fifteen? His parents have created outstanding environments for him from his day of birth to his adolescence. They make him watch great educational TV programs and read a lot of books, and they kept him in a competitive learning process, fostering his mind and brain development (Russell, 1995).

Can one say that this kid described above is more intelligent than any other kids? We do not think so; his extraordinary development is due to his parents' intellectual orientation and his learning motivation. Any other child could be developed the same way if they have been given the same environmental learning opportunities. The French philosopher Rene Descartes had already contended that we have received equal good sense at birth, but we do not take the opportunity to make it work at the highest level. This worldview is confirmed by Howard Gardner in his recent book *Intelligence Reframed: Multiple Intelligences for the 21st Century*, where he describes seven types of intelligence that we all received as birthright.

If this is a normal worldview and it is accepted as truth, then there are no individual differences at birth regardless of race, color, and nationality of origin. Individual differences exist only when it comes to applying the received reason as birthright. Therefore, individuals who grow up in the same climate and who have the same amount and quality motivation for learning will likely succeed equally. Differentiation can exist only based on self-motivation related to how much time each individual disposes to respond to the learning challenges (Carroll, 1963; and Bloom, 1971). Therefore, individuals who spend more time in his or her study, watch more educational TV programs, and read more books can make a difference in a given test, and his or her intelligence quotient can appear to be superior (Carroll, 1963; Bloom, 1971; and Russell, 1979) even though intelligence is stable thing inside.

It is quite apparent that an individual is able to make that difference only in his or her domain because it is impossible to pursue two different fields and succeed in both at the same level if they are not related to each other. It would be necessary that he or she prioritizes one field and neglects the other one in order to spend adequate time in one of them and to be highly good at it and successful in it. Otherwise, there will be neither mastery nor specialization in any respect.

We hope that Gardner referred to the degree of the individual's capacity to deal with abstraction when he spoke about logical-mathematical intelligence in his book "*Intelligence Reframed: Multiple Intelligences for the 21st Century*", which could be applied to any field of knowledge. This is the reason why that intelligence is used in education. It can also support why a student who makes a lot of progress in one field or course may encounter difficulties working well in some other fields or areas of study.

The grounds to succeed in a given field are in one's mind and brain, but the person needs to choose that field and respond to its challenges satisfactorily. The brain, however, should be developed in all areas where learning takes place, responding seriously to challenges that come with it emerging from the environment. Fortunately, an individual's brain has limitless potentials (Russell, 1979) for doing so as long as and as much as the person seeks it and deploys necessary efforts to make it happen. This is why parents and teachers should create rich learning environments and motivate the student at an extraordinary level to shape their mind and modify their brain in order to stick with learning all the times (Jensen, 2005).

Research findings reveal that for each activity a person undertakes, he or she creates a gold occasion for the brain to develop itself, and so does that person's mind. It is not different for the baby who's moving in his or her environment. When the baby looks, hears, touches, smells, or brings an object to his or her mouth, those senses get connected to their respective parts of the brain, and the more the baby does them, the more those connections become firm permanently as long as they are in use. As the environment is a key to learning, we urge parents and educators to offer their babies or students extraordinary learning environments. Mothers should start creating this outstanding environment in the course of their pregnancy because babies could turn out to be stars and gifted students, starting in their moms' bellies (Dodge and Heroman, 1999).

For example, they need a sincere guide in the study of the complexion of their body and the purpose of each organ, especially the reproductive ones. They are critically in need for this aspect because ages sixteen to seventeen are of craziness, foolishness and thoughtlessness. They also need guidance when it comes to studying humanities and philosophy, where they are seeking answers to all metaphysical and existential questions. Kids need parents and the school's guidance, especially at their young ages, when they engage in the investigation of themselves while acquiring different knowledge in different domains. They can be lost and be filled up of lust for the loss of their parents and the great society, become disillusioned, and disoriented if there is no guidance to help them answer the most difficult life questions.

Youth and Identity Crisis

How could youths survive an identity crisis (Erick Erickson, 1968) experienced from ages thirteen to eighteen? In this period, young people don't know what is good for them and can barely differentiate right from wrong, and they are mostly wrong; they make a lot of mistakes, which sometimes cost them their lives. This is the reason why they need direction, a counselor, an advisor, or a mentor that they may not be able to find at home. Anyhow, they need some experienced or educated people to help them in the midst of their wrongdoings and to impose a psychosocial moratorium, which Erickson (1968) thinks is "of the utmost importance for the process of identity formation" (p. 157). Parents and educators who fail to use their moratorium power hinder this process or mess up with it.

It is necessary that institutional instructors and educators create curricula that young people can use as solutions to their identity crises. It would be dangerous for the society if those young people enter adulthood with unclear identities. It is even more dangerous when some of them rush that process immaturely and get lost and jeopardize the rest of their lives for no paradise is out there while eating chicken and French fries with no wise thoughts.

As this educational research does not propose to go beyond young people's educational needs, we prefer to stop speculating on the importance of education in general here. We want instead to start thinking about the process of getting young people where they have to be educationally, socially, and culturally on the basis of teaching them what they should know in order to promote them to the next level.

In order for a student to be promoted to the next level, he or she has to successfully pass what educators call a summative evaluation, or the final test, which could be internal or external. A summative test is internal if it is administered by the teacher or the school administrator; it is external when it is given by the state, such as the Florida Comprehensive Assessment Test (FCAT).

Summative evaluation is performed by a teacher at the end of his or her course or the end of the year to determine if a student can be promoted or not. It means that summative assessment is the last assessment of the session. Before the final assessment, it may be possible for the teachers to perform a number of ungraded formative assessments whose objective is to thoroughly control students' level of understanding. Summative tests are also administered by the state to determine if the students can be finally promoted. A state test is considered as the toughest test ever. The state test called FCAT or Florida Comprehensive Assessment Test is considered as Florida's effort to increase student achievement.

Educators are very concerned with their students' performance on a state test. The results of that test put a lot of pressures on educators, the principals especially. This is what Linda Perlstein (2007) expressed in her book *Tested: One American School Struggles to Make the Grade*. She talks especially about McKnight, principal of Tyler Heights Elementary, one of Maryland's high schools. Perlstein (2007) reported that the principal was very anxious about the test results for her third, fourth, and fifth graders, who took the state test in reading and math.

According to Perlstein (2007), the principal could not position herself; she might have lost her mind when she checked the empty mailbox. She could not wait anymore for all the results had to be published and laid at the principal's desk, which was not effective. The state test result is the determinant of a child's failure or success in school and perhaps in life in general. One can understand the principal's anxiety.

Sometime after, in the course of the same day, somebody came and gave a package containing positive exams results to principal McKnight. It was exaltation, joy, and excitement that could not be kept inside. The principal screamed and high-fived the teachers; the students danced, and the teachers exhaled deliverance and triumphant breaths (McKnight, 2007). What would it be if the school failed the state exams?

Student achievement is the main purpose of schools, is the parents' expectations, and the country's dreams. In order for the students to be successful, they must be educated and tested properly using the highest instructional technologies, which can lead to success. Technology includes teaching and testing methods, which will be the focus of the research.

Purpose of the Research Study

The purpose of the research study is to develop an interesting method of testing that can be a great technological tool in the educators' hands at all levels: elementary and high schools, colleges and universities, and state and national levels. However, the public schools are mostly targeted. This technique is called bell-shaped testing (BST) commenting Benjamin Bloom's *Taxonomy of Educational Objectives: Cognitive Domain*.

There will be a corollary purpose related to the first one, which is a teaching method whose objective is to support the method of testing. Without the latter, the former would fail to reach its goal. Unfair treatment would be given to the poor and innocent kids and whoever has to go through any formal institutional and educational system.

Methodology

It will be necessary, beforehand, to conduct a needs assessment, which may be merged up with some kind of literature review to know if the proposed method can be better off to save the educational system, which appears to be on the precipice, the cusp of fall, unknowingly. The results of this research are going to be a wake-up call.

In order to arrive at this stage, it will be also necessary to conduct an empirical research based on which we may develop a questionnaire with open-ended questions to collect some participants' opinions. It means that we will collect some data, analyze them, and come up with valuable conclusion after searching for data validation and trustworthiness based on a qualitative research method.

Lastly, the research will have to be evaluated to ensure it was well conducted by using an evaluative approach. However, this method will be used in this section only. The qualitative method and the content, input, process, product method are integral parts of our main research method, labeled NAPOQMER, standing for "needs assessment, planning, organizing, qualitative or quantitative method, evaluating, and report." Before we get to these points, it is critical that we present and reflect on the actual assessment methods used by the teachers and school leaders, commonly called traditional methods of testing, to be enlarged by the Bell Shape Testing System (BSTS).

CHAPTER I

Traditional and Actual Methods of Testing

We use the expression traditional methods of testing here to refer to formative and summative assessments. They have been used traditionally by educators to measure the students' understanding and achievement (Gardner, 1999) in order to review the previous lessons taught if necessary or to evaluate the method previously used by the teachers. They have also used those traditional assessment methods just to decide if a student can be promoted to the next level. To be more specific and detail-oriented about those testing methods, it is necessary to briefly present them separately.

a. Formative Assessment

Formative assessment is a type of assessment that teachers practice in the classroom to evaluate and gauge their students over time and improve learning (Robert Marzano, 2006; Thomas Guskey, 2007; Dylan Wiliam, 2007; and Susan Brookhart, 2010). Brookhart (2010) goes a little further to add that teachers and educators use formative assessment to plan lessons and rubrics.

Formative assessment is considered a learning tool to help teachers know what kind of feedback they should provide to the students, whether they are successful or not on the last test. Excellent teachers should acknowledge students' efforts when they do well, very well, and excellently, and if there should be an innovative way of doing that more efficiently; they should take advantage of feedback time to improve and to pass their summative tests. It is also the dearest moment to correct wrong answers with intelligent explanation, to enhance students' understanding and enlighten their path toward success. Lastly, it is a way to know if the method used was efficient and sufficient, a way to see if state-standardized teaching method requirements have been successfully followed.

b. Summative Assessment

Summative assessment is the students' evaluation administered by the teacher or any authority that has right to pronounce that the students fail or pass the class, or should be promoted to the next level. In order to take such a decision toward the students, one must be sure about their backgrounds, what they have learnt during a certain period of time where an appropriate curriculum has been taught based on some established standardized content and teaching criteria.

Certainly, that person vested with such authority has the necessary data to conduct the test and to evaluate the students' performance and publish the results. At the last result, one can think that everything would have been well planned as teamwork is composed of the Florida School Board members in the testing department and the school leaders. However, it seems that there is not too much of interdependence and that the testing department does not provide information to the school leaders in time manner which could be a problem for students' achievement (Guskey, 2007).

Differences between Formative Assessment and Summative Assessment

Formative assessment is administered by the teachers in the classroom to measure the students' understanding of what has been taught in order to foster their understanding. It is also administered to prepare the summative assessment to emphasize a method of teaching or to modify it, to keep teaching the planned curriculum, or to refocus the learning goals and objectives.

Summative assessment evaluates the student's accumulative knowledge and skills for a period of time. It does it as final judgment. The state does it once, but the schools and the teachers may do it several times (two to three times per semester or per year) as instruction control and a way to help student with accumulating grades. Unfortunately, some teachers administer it only once. Burke (2010) thinks that all the teachers need to "build a repertoire of both formative and summative strategies to help all students learn" (p. 3). Only at this time that one can say that formative assessment meets summative assessment.

It is critical to think about the state final exam, identified as summative assessment, because such exam should not be administered once so kids would not miss the super-train forever (Burke, 2010). If it's so, summative assessment should be well built in such a way it is not considered as a kind of punishment in the student's eyes. This exam should be fairly and democratically organized. In order to be well and fairly organized, it should be in the interest of the exam taker. It means that the exam should be taken with pleasure, but not under stress, and the exam content should be based on what has been taught in the classroom.

Because the assessments have been unfairly administered, the students have been struggling to pass them and to enjoy their learning time. Some kids are afraid of school; they hate school because school is not a pleasure, but painful; young people quit school and they dropout and develop some kind of truancy and they become delinquent. There is a big concern here which demands our thinking so we can offer some assessment alternative. **Bell-shaped testing**, which will be defined and developed with details later after we perform the needs assessment, is that alternative we are talking about being the focus of our research. It is not an alternative to the existing summative assessment, but a technique of structuring that test fitting the curriculum, and balancing and distributing the questions related to the number of questions asked and their position in the scaling nutshell.

Before we go too far, it is extremely critical that we think about the meaning of success in school not only for kids and young people, adolescents, but also for regular and returning adults. Testing is not only kids' concerns but also young adults' who take back the road that they left at some particular time because of personal reasons that we can't even explain. Now, educators are talking about andragogy, method of teaching adults because they learn differently comparing to kids and youth learning (Knowles and Holton, 1998). They will need to be assessed too. Let's make them come back with pleasure and free themselves from life struggles and economic cornucopia.

CHAPTER II

A Brief Overview of Some Educational Theorists on Assessment

In his interesting research on "Content then Process: Teacher Learning Communities in the Service of Formative Assessment" presented in a collective research, *Ahead of the Curve* edited by Douglas Reeves, Dylan Wiliam presented a Pleiades advantages of success in school or on a particular test. He pointed out that the benefits are common for the individual being successful socially, educationally, politically, culturally, and economically.

Besides Dylan Wiliam, many other educators and writers advocate for students' success in school and make it clear that formative assessment is the way to go and get the students up to high achievement. Among those theorists figured: Marzano (2001, 2007 and 2010), Brookhart (2010), Stiggins (2007), Guskey (2007), Ainsworth (2007), Popham (2003), Burke (2010), Reeves (2002 and 2004) etc. Some others criticized the strategies used by the state to assess the students once a year (Reeves, 2007 and English, 2010).

Wiliam (2007) is one of the greatest advocates of classroom formative assessment practice. He connects this practice with success of the students' achievement in life and positive impact it can have on the entire society and different aspects of life. Wiliam (2007) thought that the level of an individual's success in school matches his or her level of success in life at all respects. He wrote himself the following: "If you achieve a higher level, you live longer, are healthier, and earn more money. For those with only a high school diploma, the standard of living in the United States is lower today than it was in 1975; for those who have degrees, it is 25 to 50% higher" (p. 183).

Therefore, it is necessary to raise students' achievement from one level of standard deviation to another one to change not only their lives, social and economic status as individuals, but also the country's economy and its inhabitants' level of living. And this change is due to classroom regular use of assessment, which has potentials of raising students' achievement by 0.4–0.7 standard deviation. This achievement can also "raise the United States into the top five countries in the international ranking for math achievement" (p. 189).

Marzano (2001, 2007, and 2010) was concerned mostly with formative assessment frequency found related to student's achievement. He referred to results of other theorists' researches in that domain (Robert Bangert Drowns, James Kulik, and Chen-Lin Kulik, 1991). He reported that they have conducted twenty-nine studies on repetitive assessment and all of them demonstrated that the more students are formatively assessed the more they are able to achieve at the highest level.

Robert Marzano thought that teachers must start assessing the students in the beginning of the learning episode and they should keep doing it to the last breath. He again thought that teachers can use a wide variety of formats in the formative assessment process, which include: "both formal (Paper-and-pencil-quiz) and informal (a discussion with a student)" (p. 9).

Marzano agreed with other educational theorists that summative assessment is given at the end of the episode or semester, or the end of the year, and that this test is mostly administered by the state to assess not only the students, but also the teachers and the schools. This is why it is so important that teachers assess more often the student formatively in order to be successful on the state assessment as practice made perfect.

Brookhart (2010) expressed similar concerns with other educational theorists including Robert Marzano, especially in testing frequency. But she went a little bit further to highlight the importance of higher-order thinking assessment which consists of assessing students' understanding at different levels. She did not use the word frequency, but "regularly" instead. According to her, it is critical to design assessment that shows students' thinking. It is even more critical to assess higher-order thinking. This way of testing is a way of increasing learning. She added that using higher-order thinking assessment not only increases students' achievement, but it also increases students' motivation. When students are motivated, they virtually learn at an exponential pace.

How often a teacher should perform higher-order- thinking assessment? Susan M. Brookhart is not very clear about it as Marzano does imposing assessment as soon as the class begins to the end of the semester. Nevertheless, it seems that Susan Brookhart opts for weekly assessment, which may create some equilibrium in the students' learning process toward highest level of learning development for a highest level of achievement which is the expectation of all educators.

More precisely, Brookhart would like to know educators' expected success dimension. She asked the following question: "How large can we expect this effect to be?" (p. 8). To answer this question, she referred to Higgins, Hall, Baumfield, and Moseley (2005)'s research. Those researchers conducted twenty-nine studies on verbal and nonverbal reasoning tests, and they found 0.62 outcomes. The same researchers conducted other studies on reading, math, and science test, and they found 0.62 on achievement of curricular outcomes. According to Brookhart, 0.62 is sufficient to move the students' achievement from the 50th percentile to the 73rd percentile on a standardized measure (p. 9).

Therefore, it is always good and recommended to conduct classroom assessment very often (Marzano, 2001). Moreover, teachers who want their student achieve at the highest level, should not only test the student often, but they should also check for higher order thinking (Brookhart, 2010).

Stiggins (2007) also advocated for formative assessment that could foster the students' learning. To him, it is impossible for students to be successful in school and to pass the state exams if they are not assessed very often in the classroom. This is the basics of all instructions and learning, but the deepest thing to consider here is the strategies educators should utilize to assess the students and make the tests not having objective of failing them. In order to assess the students effectively, teachers and educators must construct quality assessments.

Rick Stiggins thought that the assessor should primarily have a clear purpose, a sense of why he or she is assessing the students. In addition, he or she must target a clear achievement. Some other practitioners think that it is very critical for kids to participate in that process. Students should have a sense of constructing their success so they have the idea of achievement in their head since the beginning. Next, according to Stiggins (2007), the teacher-assessor must design an assessment that is based upon the assessment's purpose. Lastly, assessment should be done accurately and the result needs to be communicated effectively. Otherwise, there cannot be any progress or great achievement, not only in the end-of-course assessments, but also in the state or national assessments (Stiggins, 2007; Popham, 2008) considered as summative assessments, those administered only for final decision.

James Popham is convinced and has convinced educators and teachers that there is no greater action to be undertaken to save the students from external hard assessments than practicing standardized classroom formative assessments. Popham (2008) presented himself as one of the greatest and lavished formative assessment advocates the educational system has ever known. Here what he wrote: "If you're looking for an advocate of formative assessment clearly you've come to the right place, given the topic of this book, it shouldn't surprise you that I am one" (p. 13).

James Popham repetitively urged teachers to use and practice formative assessment not only to put their teaching on the wheel of progress, but also to keep the students awake in their learning process. It seems that he posed it as fundamental that the students also should assess themselves and put themselves in the progress track. His exceptional view expressed in his book *Transformative Assessment* (2008) comes from the Council of Chief State School Officers' definition of formative assessment:

> Formative assessment is a process used by teachers and students during instruction that provides feedback to adjust ongoing teaching and learning to improve students' achievement of intended instructional outcomes. (p. 5)

Even though Popham (2008) expressed his profound admiration for the Council's definition, he proposed a new one which is not too different, but a more precise definition of formative assessment. This new definition puts emphasis on the following points: planning, identification of student status, and adjustment. To him, adjustment should be the aim of administration of formative assessment during the process; it will make sure that the student gets the desired goal of the teacher's instruction and their learning. But in order for adjustment to be possible, teachers must know the students' status so progress can be visualized, planned, and be remarkable when time comes. Based on those considerations, it seems that Popham (2007) provided the must finished and exhaustive definition of formative assessment in history. Here is that definition:

> Formative assessment is a planned process in which assessment-elicited evidence of students' status is used by teachers to adjust their ongoing instructional procedures or by students to adjust their current learning tactics (p. 6).

James Popham furthered to encourage school administrators to participate in some external formative assessments organized by statewide assessment centers that "hitch a profit-making ride" (p. 9). He mentioned one of them that conducted an interim or benchmark assessment every year. He thought that school participation in this kind of scholar activities could be a plus for students' achievement. Yes, it is a great means of making leaning progress as long as the results will help teacher make adjustment in their teaching and that students will do the same in their learning process. Any test that is administered to help make final decision must be classified in the category of formative assessment, but not summative assessment.

Those benchmarking tests had to be organized more often, or monthly or quarterly. In addition, why there isn't many other external testing centers, whether they are statewide or countywide or citywide? It would work better keeping the students busy. The more students practice formative assessment, the more it is possible for them to reach a higher achievement (Marzano, 2001).

Popham (2008) has an excellent view of instruction and formative assessment. Interestingly, he accepted that he borrowed this view from Paul Black and Dylan Wiliam, pioneers of formative assessment theory. They published, in 1998, their research report on this matter under the following title "Inside the Black Box: Raising Standards through Classroom Assessment." As one can read it, they were looking for standards, which can be accomplished only in formative assessment. In other words, there is a unique way to attain standards, and that unique way is formative assessment that needs to be practiced in the classroom by both teachers and students.

Like Popham (2008), Black and Wiliam (1998) thought that it is critical for teachers to know their students' skills and understanding level, to teach them curricular lessons, to administer frequent formative assessments, to provide the results, and to provide feedback to them. After that, teachers should adjust their teaching method and make sure they do not leave any students behind.

Black and Wiliam (1998) made three assumptions which they tried to verify based on a meta-analysis conducted on more than one thousand book chapters and professional articles on assessment. Before pointing out the researchers' report, it is good to say that they viewed assessment as simple activities undertaken by teachers and students, and it is used for feedback and re-adaptation, adjustment. This assessment is called "formative assessment." In order for this assessment to be standardized,

There should be evidence that formative assessment raises standards.
There should be evidence that there is room for improvement, and
There should be evidence about how to improve formative assessment.

Fortunately, those assumptions were verified by The Black and Wiliam (1998)'s report. Therefore, it is proven that they had highly contributed to the instructional and learning standards. However, one can doubt of them being pioneers of formative assessment after the reading of Linda Allal and Lucie Mottier Lopez's text from the University of Geneva twenty-five year ago, "Formative Assessment of Learning: A Review of Publication in French" and Maddalena Taras's "Assessment-Summative and Formative: Some Theoretical Reflection," published in 2005. All of them referred to Scriven (1967) and Bloom (1968 and 1971) as two real pioneers of formative and summative assessment theory. We are not going to present Mantz Yorke's view here, and we want to postpone Taras's text because she has a different view promoting, instead, summative assessment upon formative assessment. It is unusual which makes her the index of the fingers and which deserves our close attention.

Burke (2010) started with idea that formative assessment gets attention of all the educators because of Bush's 2001 legislation on education "No Child Left Behind." Based on this legislation, state tests were created to impose adequate yearly progress for each student in order to be promoted. If majority of students do not progress in a particular school, it is considered as a failure and it may be difficult for it to receive some kind of grants. Therefore, teachers should not wait until the end to gauge their students; they had to assess them very often during the course of their instructional sessions in order to give them feedback and adjust their teaching method to assure they understand and make progress. According to Kay Burke, the unique way to that is practice formative assessment.

Burke (2010) enumerated four kinds of assessment, including practice benchmark tests, interim assessment, short-cycle assessment, and end-ofquarter tests. She was not so clear about short-cycle assessment because she didn't define it, but she presented an assessment cycle containing eight steps that she called common assessment cycle. It is important that we present her common assessment cycle found in her book titled *Balanced Assessment from Formative to Summative* (p. 29). It is better to report Burke's "Common Assessment Cycle here as it is presented in her book:

A Common Assessment Cycle,

Step 1- Identify the power standards in all content areas, K-12.

Step 2- Repack the standards and target the big ideas and LOTS (language of the Standards).

Step 3- Write test bank questions to assess knowledge and skills that can be judged objectively.

Step 4- Create performance tasks, checklists, and rubrics to assess work that requires Subjective judgments.

Step 5- Teach the standards and use diagnostic and formative assessments to provide Feedback.

Step 6- Examine student work and analyze data to determine what needs more work.

Step 7- Different instruction and use additional formative assessments to provide Feedback targeted to the standards.

Step 8- Administer and evaluate summative assessments and assign final course Grades.

Burke (2010, P. 29; Fig. 3.1)

Douglas Fisher and Nancy Frey (2007) also came with a new and innovative way of communicating the importance of formative assessment. It is a search for ultimate standards without which there is no real teaching and learning. Fisher and Frey (2007) used a common word that is extremely important in learning that teachers can't dare ignore. This word is simply understanding.

To Fisher and Frey (2007), understanding is based on the quality teaching, learning, and on the targeted point. If the students don't understand what was taught, they will not be able to be successful on any summative assessment; the teacher and the students' time have been wasted. Therefore, it is necessary that the students understand the lesson and show their understanding. As they will not tell and demonstrate it themselves, teachers should check personally in order to ensure that learning is in progress. Fisher and Frey (2007) thought that teachers should do it by means of formative assessment followed by feedback to know where they are, by feed up or feed forward to look at the future because learning is about future where assessors will decide if a student should be retained or promoted.

The objective of feedback is to help the students to be aware of their lack, misunderstanding while they pretend that they understand, which is not good for them. Sometimes, they are aware of it but they would not state it. Therefore, teachers need to check their students' understanding by asking questions, checking their languages, their skills and backgrounds and so on. Teachers use those stuffs as types of formative assessment, according to Fisher and Frey (2007). Some other formative assessments could be quizzes administered weekly at least so feedback can be provided to the students in order to adjust teachers' teaching and students' learning. It's only this way teacher can feed forward to help students reach greater achievement on external or summative assessment.

By this time one has to have in mind that Summative assessment is administered in the end of the program to judge the students' skills and understanding. Why should teachers wait for the last day to know what their students do when it is possible to know it before in order to help them do better? One way to do it is practice formative assessment, which is an ongoing process during classroom time (Fisher and Frey, 2007).

In our endeavors to conduct this brief literature, we would like to divide practitioners and theorists, who are so concerned with assessing the students and care about their achievement, in two categories: a category of theorists who are formative assessment advocates and another category who express their concerns about how educational officials organize the final tests considered as summative assessment. Among them figured Larry Ainsworth, Douglas Reeves, and Fenwick English. We think that there is more of this kind, but we just call upon those who are on our reach, and we hope that some other practitioners reveal them to the forthcoming generation. Now, it is good to briefly present those theorists' views of summative assessment mentioned above individually.

Ainsworth (2007) criticized the educational leaders at state level. To him, assessments are not built to benefit the students. Therefore, it is clear that teachers, school administrators and parents are not satisfied at any respects. Ainsworth (2007) highlighted the following negative aspects regarding the organization of the state tests: isolation of kids, result does not create any impact on the child's learning growth, time elapse creates relevant default, state focuses on groups of students instead of individuals, lack of feedback on state assessment, problem identifying "what students know and are able to do" (p. 82).

Larry Ainsworth thought that it is necessary to practice classroom formative assessment so teacher could create room to provide feedback to the students. For there is no teaching and learning without feedback, and feedback is empty of sense if there was no assessment. Therefore, this necessary assessment should be formative by nature being not a judgment, not for grade to retain or promote any students, to decide if the teacher should feedback or feed forward. However, in order for this assessment to be formative by nature, it should be administered during the class time, but not in the end of the project.

Ainsworth (2007) thought, as a fact, that state summative assessment does not help at all the students, but it confines them to a certain state of burden. He remarked that the state test results come very late, and this is the reason why summative assessments alone cannot help the students. It is a kind of punishment given to kids (Reeves, 2007; English, 2010). State test needs to be preceded by formative assessment in the classroom whose goal is to prepare the students for it so there should be no surprise on their performance. One thing is certain: "Common formative assessments produce results" (Ainsworth, 2007, p. 99). This is what would make instruction amazing.

Reeves (2007) and English (2010) criticized the external testing a little deeper by making educational state officials responsible for the students' failure. Douglas Reeves was dreaming for a free-test day to come one day. State's summative assessment is stressing and oppressing them up. They will never be able to make up for they fall in the left side of the bell curve negatively and unknowingly created by state assessors.

According to Reeves (2007), educational state official created a bell curve which they divided into two parts or groups with poor students on the left side of the curve and rich students on its right side. The assessments are prepared to make it happen and favor rich people. This division is not good for students who come from parents with economic situations, who are powerless. Therefore, there is a kind of inequities that should be eliminated in order to render justice to a class of poor, low economic status and powerless, which imposes a superimposed educational structure. English (2010) would provide a clear explanation to this traditional problem by making people know that those who are testing the students are those who have power, those who are the policymakers, who are imposing the curricula, who are writing the textbooks, and the like. They want their kids to be triumphant over the other group already named above.

Anyway, Douglas Reeves requested solution for the inequity problem that sells out the educational standards. Reeves (2010) posed another crucial problem that needs to see with educational leadership. "Assessment is a leadership issue," wrote Douglas Reeves. Posing educational leadership problem is posing the problem of ethics and morale. The presence of inequity indicates ipso facto that there is a lack of leadership. To solve this problem, educational leaders should assess kids sincerely. Reeves (2007) referred to Parker Palmer in his book *The Courage to Teach* published in 1978, to state that courage requires to lead, to mean that educational leaders who are administering state assessment are not leading. Therefore they are those who are creating what he called inequities.

English (2010) assumed that these officials were aware of their fault when they are undertaking actions and deploying efforts to remove the system from its mire. Otherwise, there would be no governmental educational strategies at the Federal levels which are justified by Ronald Reagan's "A Nation at Risk" in 1983, George W. Bush's administration's "No Child Left Behind" in 2001, and the Obama administration's "Race to the Top." The first law just recognized that there was something that was not working in the educational system, but they could not locate where it was in order to fix it. However, Bush government has discovered it and brought solution to it by its "No Child Left Behind." How had they left child behind? What category of children they left behind? Douglas reeves identified them in his inequity analysis. Barak Obama did not need to search for that problem and for solution anymore because it was found and solved by Bush. Everyone is aware of efforts that have been deployed to work with all types of students (students with economic problems and disability needs). It is normal that the actual government comes with the slogan "Race to the Top."

This progression in the governments' interventions was effective because education, after 1950, became a national concern. Someone just came to fix the problem of educational marginalization that has been created by traditional governments inherited, maybe, from British government, according to Fenwick W, English. He affirmed that the pilgrims traveled with it in their mind, unconsciously, maybe. They just had a sacrosanct belief, which Michael Parenti called *sacranda*, a mental value that makes some people understand that they should compose a dominant group, and this dominance spirit, unfortunately has affected the educational system. Enslaving someone in his spirit and you will be able to enslave him or her in his or her body. This is why the problem starts in the educational system. Now one should ask what kind of education students are receiving from the authorities.

English (2010) seemed hit the target when he posed this heavy heritage as mental situation that has put the entire nation at risk. It is when we consider that those kids with tortured mind are called to become the nation's tomorrow's teachers, principals, school administrators, assessors, superintendents, House of Representative leaders, Senators, governors, and presidents. What kind of leader could they become with this shameful mentality?

To address the assessment problem, Fenwick English thought that assessors must be sure they design assessments that measure relevant aspects of the established curriculum imposed to the schools. Unfortunately, English (2010) discovered that there is no such curriculum that has been taught to the students because the textbooks have, for long, replaced the curricula. Therefore, the educational system assesses kids not based on the curricula, but based on the textbooks. This is what needs to be corrected and fixed forever. To understand the system, one has to ask questions such as: Who writes the textbooks? Who imposes them? Who are making money with the textbooks? Who prepares the exams? Do the people who prepare the exams have their own kids learning in the system? Is it possible for their kids to know in advance the exams (English, 2010)? It is remarked "the ruling class represents their interests" (p. 7), and they have every authority's blessings. Here is what English (2010) borrowed from Parenti (1978) and wrote,

> The interest in an economically dominant class never stands naked. They are enshrouded in the flag, fortified by the law, protected by the police, nurtured by the media, taught by the schools, and blessed by the church. (p. 7)

English (2010) thought that the test should not be hided from only some classes of poor and colored people and other minorities while it is open to some others from the dominant class. It is unfair and unjust; it's a leadership problem as Ainsworth (2007) discovered it. This is the source of educational inequalities. English (2010) again thought "that we should initiate the process of untangling this racist, sexist, and class-based system masquerading as meritocracy is long overdue" (p. 4). We should eliminate the hidden curriculum that has never been taught. This hidden curriculum approved by the government is called by Bourdieu and Passeron cultural arbitrary (English, 2010). Education is a public business; this is why all the people have right to participate in its management and to know the truth about curriculum and everything that must contribute to the success of their children (Dewey, 2012)

The real solution, according to Fenwick English, is laid in alignment of the curriculum with teaching and learning, and testing. Teachers must be able to teach honestly the curriculum and the test so the students can be prepared for greater achievements. Otherwise, their failure is not theirs, but the educational leaders' because educational leaders and assessors are measuring the students' skills and abilities in false.

To end this chapter, we think that it is necessary to, at least, expose ideas of one theorist who gives more value to summative assessment to the detriment of formative assessment. That theorist is named Maddalena

Taras. She published, in 2005, an article in the Journal of Educational Studies titled "Assessment-Summative and Formative-Some Theoretical Reflections" in which she held that summative assessment is the basis and departing point of all assessments. Taras (2005) recognized that there exists another assessment that has some values, in citing other theorists such as Scriven (1971) and Broadfoot (2000) to support her view about FA values subordinated to that of SA.

Taras (2005) referred to Scriven (1971) to declare that "formative assessment is the same process as summative assessment" (p. 468). How could they be the same when the former is administered daily and the latter at the end of semester or the year? What about checking for understanding during and after each presentation (quizzes or oral questioning), or weekly or monthly (test on a chapter)? They are not final assessments, and they maybe not graded, or they represent a small percentage of the entire grade. Those assessments are called formative assessments, but not summative assessments, which are administered at the end for decision of final grade. Formative assessments are administered to provide feedback to students so they can improve and reach a greater achievement on the final, called summative assessment when it is well forecasted (Bloom, Hastings, and Madaus, 1971).

How summative assessment or evaluation in Bloom's vocabulary can be forecasted? Benjamin Bloom, Thomas Hastings, and George Madaus intended that teachers and students are able to predict the outcomes of the last or final exam weather it is organized by the schools or by the states, or by the national assessors if they prepared it using formative assessments. Here is one more reason why one cannot consider formative assessment as the only existing assessment. It is the teacher's role to help the students to master the notions that are included in the curriculum aligned to the testing system. They called this method "mastery learning" and every time people make this reference, Bloom's thought is in evidence. This thought is exposed in his famous book published, in 1971, with two other cowriters cited above. The title of that book is *Handbook on Formative and Summative Evaluation of Student Learning*.

Whether it is formative or summative assessment, students could reach the greatest achievement ever if assessments were made based on what we call "bell-shaped testing method" (BSTM). This is a method based on which the assessor uses Benjamin Bloom's educational objectives to ask questions ordered from simple questions to complex ones. We mean from more simple questions to less simples ones and from less complex questions to more complex ones.

Simple questions are memory questions. There are different memory questions that assessors can ask, including episodic memory, semantic memory, emotional memory, automatic memory, procedural memory, shortterm and long-term memories (Tileston, 2004). There are also different types of complex questions (ones that check for skills and understanding). Bloom's levels of understanding can teach us about how question can be complex. We will be more explicit later. We will see, in the following paragraphs, how this system is useful and beneficial for the teachers and the students.

In order to go deeper in that educational research, it is good and the perfect moment to introduce our working method which is NAPOQMER (needs assessment, planning, organizing, qualitative or quantitative method, evaluating, and reporting).

Chapter III

Needs Assessment Analysis

According to Blanchard and Thacker (2010), analysis of a system is made when that system is not working properly in order to search for the causes and solutions to the problem. They were talking about organizations that are not successful because of employees' lack of training. They recognized that not all employees need training, and in order to provide training to those who need it, they advise managers to conduct training need analysis (TNA). They concluded that those who need training should receive it in order for the system to do well and function successfully, but those who are not ready for training while they are not able to produce based on the organization's expectation should be kept out because it could be a waste of time. It is not different for any other systems such as educational system, more precisely the testing system. This is what needs to be analyzed now in order to change the instructional and testing methods educational leaders and assessors have been using to instruct and assess the students, which seem to be a failure. This activity can be undertaken in any educational system at all levels, including technical schools, colleges and universities.

There is no difference between analysis and assessment, in this context, even though the latter is larger. Anyway when you assess, you analyze, and when you analyze, you evaluate to identify what needs change or not. This is what we propose to do, right now, in this research: needs assessment analysis. In order to do so, it is critical that we analyze the actual system of assessment, its results, students' needs in this area, teachers' needs in this area, and make the conclusion on this section. From there, it will be necessary to consider if the testing system would need a new method of testing.

We want to be pragmatic conducting research related to real problems with which the population is confronted and needs solutions. We hope that we will make real experiences and find something that is really "relevant in day-to-day practice of educators" (Biesta and Burbules, 2003, p. 1). John Dewey, in his educational postmodernist approach, thinks that, not only we have to be pragmatic, but also to be able to make progress in our educational research. This is why it is necessary to build upon the established system that needs to be updated. We are investigating to acquire knowledge that can lead our actions and activities, borrowing the expression of Biesta and Burbules (2003).

a. Facts in the Actual Testing System

In this educational system, teachers and students are gathered officially in classrooms to transfer knowledge and to learn based on established standardized curricula. In the end of the learning process, educational leaders provide official tests to the students and make major decisions related to their achievement or failure.

External assessments or summative assessments are administered once a while. They are made to judge the students and decide if they fail or pass. Sometimes, students are allowed to retake the exam if they fail. After that the decision may be final.

According to Douglas Reeves, the questions are more likely to be memory-type. For some other practitioners, questions used to be understanding-type only. It would have been better to have mixed memoryunderstanding questions arranged in such a way student can use their memory and their skills in the same time (Bloom, 1956, Marzano, 2001 and Wiggins and McTighe, 2011).

Assessors take long before they publish the assessments' results. This is why Reeves (2007) poses the problem of time. It means that publication of the test results is not made in timely manner.

State assessments divide the students into two broad groups: Rich students and poor students (the minorities), according to Reeves (2007) who mentions the problem of the bell curve. This situation creates the idea of competition instead of cooperation, which is not good for the learning process.

There is also a problem of assessment standards that is mentioned by Reeves (2004) and by English (2010). They thought that school officials have a hidden curriculum, and that test questions are not taken from the working and open curriculum, but the hidden one. It is a kind of abuse.

If what those practitioners declare regarding this case is true, it would be a moral or an ethical problem, a leadership problem (Reeves, 2007). This will not be stopped as long as education is fully led by politicians, but not educational leaders (Dewey, 2012), and as long as "knowledge remains a source of those in power" (English, 2010, p. 8).

In the classroom, teachers administer local assessments called formative assessment to help the students perform better on the state tests. They give them quizzes, check for understanding orally. Lastly, the teachers provide feedback to them, and update their method of teaching if necessary.

It does not say that educational state leaders and teachers have any model of testing strategy related to questionnaire structure. However, some practitioners and theorists advise assessors about types of questions that should be on a test (Bloom, 1956; Marzano, 2001; and Wiggins and McTighe, 2011).

All those aspects indicate that there is no curriculum and testing standards. It is also a sign that there is a lack of teaching standards. Lastly, there is no alignment between the established curriculum, teaching and learning, assessment standards.

It is said that curriculum and assessment standards are supposed to be done to ease the learning process and make kids achieve at the highest level. However, it seems that the opposite is true and it is the reality that persists for long. It is imperative to assess the students' needs and their level of satisfaction.

Students' Needs and Satisfaction

Needs and expectations are two things leading to actions as it is not different in students' learning process. Students go to school because they need to get great instructions from educational leaders and teachers that are highly trained and competent for this purpose. Therefore, they expect to accumulate skills and appropriate knowledge able to operate changes in them. It is not without reason John Dewey declared that the school's mission is to transform the children so they can become men and women of tomorrow. In the end of adolescence, the need of assessment imposes to the society in order to decide if they become desired men and women to be useful to them and to it. At this particular and critical time, they want leaders to use assessment standards. To do that correctly, they need to be informed about the process so there would be no surprises.

What are some students' needs? We are going to make some assumptions and call upon some other practitioners such as Donna Walker Tileston to determine the students' needs in this particular area of instruction.

John A. Comenius, one of the founders of modern education, declared that youth demands good education and right. Sometimes, according to practitioners (Wiggins, 2007 and English, 2010), educational leaders are not willing to provide good education to the school age population. Therefore, youth needs to stand up and claims what belongs to them. It is not about a group of individuals (wealthy and political powerful people), but the entire population, including poor people, Blacks, handicapped, women, and other minorities. Dewey (2012) in his book titled *Moral Principles in Education* thinks that they have right to participate in their educational and decision making processes.

It is necessary to inductively translate these complaints into the assessment system that needs to be constructed or reconstructed in order to satisfy the students' needs. Firstly, students need to be critically ready for the internal and external tests. At what level do they want to be ready? According to Tileston (2008) they want to be ready at a point they can do things, take exams with automaticity (without conscious thoughts) (p. 51). Automaticity is possible when teachers provide them with:

a. Adequate opportunities for practice (referring to formative assessment which includes
   Classroom quizzes, classroom essays, classroom interaction, and homework).
b. Helps in self-assessment.
c. Help in metacognition practices. Competent teachers should be able to help them to do that so they do not accept all information as truth or knowledge. An epistemological dimension is necessary in this situation. True knowledge is accumulated by questioning. Please, see Donna W. Tileston in: Teaching Strategies that Prepare Students for High-Stakes Tests.

In addition, we would like to insert some assumptions based on pass experiences as former student and as an observer. Those assumptions should be considered as mere requests of the students who are voiceless still today as it was before. We assume that students would need and demand the followings:

1. Better treatments related to summative assessment whose purpose is to make final decision regarding their failure or success.

2. To be assessed based on a given taught curriculum and nothing else.
3. That teachers and state assessors should not use tricky questions.
4. Educational leaders to publish the assessment's results in timely manner and to provide explanation on how they were doing.
5. That they receive feedback from their teachers as soon as possible after assessments are processed, or throughout their teaching sessions. If it is oral feedback, then teachers should provide enough time so they can think and participate in that exceptional moment of the learning process (Black et al, 2003). In his research, Rowe (1974) reported that teachers provide only 0.9 seconds to students to respond to their questions while the time allotted had to be longer (Black, et al, 2003).
6. The assessment should be administered in such a way the test taker can be relaxed on the tests.
7. To know what kind of assessment will be used to evaluate their achievement.
8. To participate in the assessment planning process.
9. Teachers not to keep assessments in secrecy anymore so students don't guess on the tests.

Students' Needs Satisfaction

It seems that the students are very far from having satisfaction related to assessment in general, and related to the state assessment particularly. There must be some changes in the teachers' method of assessing the students. It's maybe what Black et al (2003) meant when they were conducting their research on assessment for learning. Their endeavors were to have the teachers to come together to work on creating common formative assessments.

Paul Black, Christine Harrison, Clare Lee, Bethan Marshall, and Dylan Wiliam undertook these activities in 2000s. They found many teachers adhering to that movement; they reached nine schools that provided a total of twenty-four teachers of which nineteen thought that was a wonderful idea. Those researchers wanted to improve formative assessments in the classrooms so the student could perform better.

According to Guskey (2007), teachers are still hiding their assessments in order to surprise their students. It is like they want to say: I catch you. This method of testing is not good. Students are not invited to participate in the assessment planning and they appear to be strange in that system of learning. Something needs change (Guskey, 2007, and Black et al, 2003).

It seems that some teachers do not provide feedback to the students and reinforce their teaching at a point that students' understanding be checked individually. Therefore, there is no progress, no improvement at both levels teaching and learning. Neither the students, nor the parents are satisfied. It is an educational failure from the instructors, but not from the students.

When it comes to considering the results of state summative assessment, one can observe that it is catastrophic every year, and the school cannot meet the adequate yearly progress (AYP) based on the No Child Left behind of George W. Bush. Chris Guerrieri made a statement regarding the state of Florida's failure to satisfy the students' need which is passing the state test, in 2012. He said that 57,000 kids did not pass Algebra exam. Originally, the number was 66,000 across Florida, but they had a chance to retake it, and only 32 percent of them have passed. Therefore, neither the students nor the teacher were not satisfied at all. It was and should be educational cries on the streets. If the students are not satisfied, the educational system is neither satisfied, nor the parents. Something needs change.

Bloom and his colleagues have already cried out to claim the necessity of this change. They systematically posed that the main objective of the educational system is supposed to operate change in the students as education is defined as "a process of change" (p. 8). How can they do it if they do not change themselves? Unfortunately, they have no will to do so (Reeves, 2007 and English, 2010). How do you want to change someone when he doesn't believe in you? Because teachers are the primary concerns, they had to do something to move the instruction level and take the children to the top as Obama has whished it to be. And, maybe, the only way it is possible is to assemble the teachers together to exchange ideas in order to create some kind of front-line formative assessment system to combat the common educational enemy that is called failure (Black et al, 2003).

Teachers' Needs and Satisfaction

According to Black et al (2003), teachers find themselves in a delicate situation; they want to please both the educational leaders by teaching unfitting curricula and hiding the tests away from their students. It seems that they are aware of it, but they have no power to stand up and pose it as a problem. Unfortunately, they can't do anything because most of them are corrupted and they think that hiding the tests is all right. They are extremely wrong (Wiliam, 2007; Reeves, 2007, Black et al, 2003, and English, 2010).

According to Thomas R. Guskey, school administrators and teachers have no patience to wait on the state test results which take longer than expected. They cannot wait for two major reasons. The first reason why they can't wait is that they want to know how their schools are doing. State results rank the schools as straight A, or B, or C, or F schools. Do they meet the No Child Left behind's adequate yearly progress? Nobody knows until the publication of the results. The second reasons why they cannot wait to have the results is that teacher would like to provide feedback to the students, which is the pivotal point of teaching and learning so there can be improvement.

As the educational leaders always delay the assessment results, teachers would never be satisfied and would always feel that some important parts of education are missing. In this horrible situation where expression of loss appears upon their face, teachers are misleading the students. It is impossible to end this part without quoting Guskey (2007) who wrote the followings related to teachers assessing their own students as it's said in the preceding paragraphs that the students have bad experience with the teachers' assessment: "This experience is a common one for students because many teachers still mistakenly believe that they must keep their assessments secret. As a results, students come to regard assessment as guessing games, especially from the middle grade on" (Guskey, 2007, p. 17).

Needs Assessment Conclusion

We have come to see how the students and the teachers are very far from getting satisfaction and have better teaching and learning experiences. The reasons why it is like that is that the educational system is struck at the top. It means that educational leaders who have the power to establish a reliable system of education to move the unproductive generation to a productive one have no leitmotiv and interest to do it. The system is productive when it provides great education to the children and teaches them for change, and teaches them satisfactorily.

Satisfaction, in this context, means that the students pass successfully the state tests, there were no problems in the curricula, and the curricula were aligned to classroom instructions and the assessments. In addition, satisfaction implies that the tests results have been published in timely manner and students have received feedback properly. Unfortunately, none of those conditions, criteria were met. Therefore, it is necessary to try something else which can help positively modify the test results to alleviate the students' hearts by means of stress free test. Stress free test is possible with the application of bell-shaped testing that has been introduced above.

Some teachers have deployed efforts in vain to help the educational system work well. They introduced formative assessments in the intent to prepare the students for the state summative assessments. In addition, some others conduct researches on this matter in order to better understand the problems and propose some brilliant solutions. The problem still stems. Moreover, they create teachers associations and work on the idea of common assessments to unify the divided educational system. The problem remains still.

Lastly, some practitioners think designing some kinds of teaching and testing based on which student could be taught from simple notions to complex ones and be checked for understanding should be recommended. The system definitely is not working so well. To make it work very well, students should be given tests not only based on the taught curriculum, but also based on a kind of arrangement of simple questions and complex ones to form what we call bell-shaped testing. Therefore, there is a necessity to rethink the structure of the tests whether they are formative or summative assessments.

CHAPTER IV

Planning for a Better Testing System

Before we go further, it is necessary that we define the word planning in general and point our fingers on its importance, and what it can help achieve. After this step, we can further try to explain planning and its importance in education and assessment particularly. J. M. Juran defined planning as "the activity of establishing goals, and establishing the means required to meet those goals" (Juran, 1992, p. 13). Juran contended that planning in general should be great and excellent. Those qualities applied to planning make it quality planning. Juran's quality planning is compared to what Douglas Reeves calls standards-based in educational planning when it comes to teaching and assessment.

According to Glickman, Gordon, and Ross-Gordon (2001), there cannot be quality planning if the planner doesn't know where he or she is going, doesn't have a map (a global positioning system, or GPS as guidance) to create a trajectory from one point to another one if he or she does not create a route. In addition to those, the planner would have to consider the notion of time to determine at what time it is possible to get there, and to know and acknowledge when he or she gets there.

Glickman, Gordon, and Ross-Gordon (2001) think that the map is needed and necessary if the planner is uncertain about the destination, he or she should create the route unless certainty is reached. Otherwise, planning process should be delayed, but not postponed. Something is postponed when the planner is not working on it anymore for the present moment and may not come back to the activity undertaken after a very long period. On the contrary, delay means that the planner is at work and keep working until a certain point of the project is attained.

This is that situation that prevails actually in the educational testing system that is divided into three branches: the school level is where teachers and administrators conduct formative assessment daily preceding summative assessment that is administered quarterly or at the end of the year. Another part of the testing system is state level (FCAT) where decision is taken at a higher level not only to judge the students but also the teachers and school administrators, and the last level is national level (NAEP, or national assessment of education progress) to check how states are doing (Grissmer et al., 2000).

At all three levels described above, educational leaders use what we call planning without which nothing serious would have been able to be realized. Fortunately all three have a common goal: the students' achievement.

However, they use different method-assessment based. It is necessary that they use a method to plan the route because it is the way of building certainty and to forwardly step with assurance. The school system, in general, uses Deming's total quality management (TQM) whose tools allow to monitor purposes in different school districts and focus on continuous improvement. It also allows school leaders to "benchmark and track students' process through charts and graphs" (Hoyle et al, 2005, p. 91). According to John Hoyle, Lars Bjork, Virginia Collier, and Thomas Glass (2005), educational system also uses the CIPP method of planning where CIPP stands for context evaluation, input, process, and product evaluation.

We admire the word input of the CIPP because what you put in is what you will have as output. In other words your output is dependent upon your input. How much does the system put in order to have high output with students facing high-stake tests?

Traditionally, educational leaders start with building viable curricula and establishing what they call curriculum alignment so teachers in the school districts teach at a common level to help galvanize equitable instructions. Next, educational leaders impose textbooks they feel that correspond to the current curricula. In addition, they hire competent teachers to teach at desired levels based on certification entry. Lastly, educational leaders evaluate the established taught curricula by assessing the students.

Those are the steps that fully complete the planning process without detailing sub-points leading to the accomplishment of the assessment project. Details would be the technical and procedural steps used to make the finished line. For example, school curriculum is made based on students' needs, family and environmental needs. There should be somehow to approach them and make appropriate decisions so a related curriculum can be created.

As CIPP is mostly used in program evaluation and that we do not have any program here except a proposal, we are not going to stay on it too long. We should use largely qualitative data analysis evaluation (DAE) in order to search for validity and reliability of our findings. After all, we'll be able to cautiously report the findings.

Moreover, there are extra steps that should be perceived to address the problem of alignment, evaluate the textbooks, and to prepare appropriate test questions based on the established and taught curriculum. Oops! The most important step has not been yet cited. That important step is the educational goal (EG) which should be envisioned. This is a way to indicate that educational planning is not an easy task.

Remember that first thing comes first and must be done first, according to Stephen Covey. The first thing of any project is the goal to be set. But do not forget that there is something to check out at the end of the project, and that thing is the performance, accomplishment of the goal. If the goal is not reached, the program fails. Is it what we mostly notice in the educational system when it comes to assessment?

The reason why educational leaders and teachers cannot reach their goals when it comes to assessment is not because they lack planning strategy or implementation strategy, but because only they never thought of structuring the assessments. Assessment structure is a good tool that can be used in both formative and summative assessment.

The majority assessments contain only recall and memory questions or problem solving questions, or inferential questions. Consequently, the exam becomes too easy or too difficult. Another fatal consequence that is attached to unbalanced assessment regarding choice of the question, whether they are multiple-choice questions or other types of questions, is that educational leaders and teachers are not checking all the fundamental and critical teaching elements which should be checked. They include memory, knowledge, skills, understanding or comprehension, and other elements as presented in Bloom's Taxonomy of Educational Objectives. Even though those elements were presented by the leaders and teachers, there still would be lack because there is no such presence of classification that is extremely critical for the success of the examinee.

I took the FELE (Florida Educational Leadership Exam) after some times of preparation; I saved a lot of information in my memory compartment using strategies of recalling simple element and complex one in addition to organizing my thoughts regarding certain cognitive elements that need understanding demonstration. Unfortunately, I encountered only high level of comprehensive questions. Consequently, I experienced tiredness, fatigues, and discouragement. Fear of testing knocked inevitably at my door. That was only at that time that I understood why Comenius J Amos, John Dewey, and Douglas Reeves talked about children's fears of school. Schools stop being, if they were, a center of pleasure.

In addition, I experienced another problem; it was discriminatory problem. Upon the examiners' representative saw me, they seemed to have expressed something different than expectation. It was, to me, a kind of intimidation, and the interpretation that came to my mind is that it would have been impossible to have a darkish Haitian skin in this room. I was panic and all my good senses for this highest assessment have been taken away, and I could not execute any of their instruction prior to entering the assessment room. Just think what would happen to me while answering the test questions lasting six hours and a half. Consequently, I failed all three tests which were subsequently scheduled: Leadership for Student Learning, Organizational Development, and Systems Leadership.

Questions that I asked myself were the followings: Have those tests been created for a category of people (Whites), but not for some other categories (Blacks, Haitians, Africans or Black Americans)? Why didn't assessor' representatives congratulate the hero candidates upon they open the entrance door? What do assessors intend by testing? Do they know what they are doing when setting to assess on the technicality of education? What do they do with the idea of assessing memory? Do they have any psychological notion of relaxation related to testing? Those questions need to be answered before creation of the next assessment. Being pragmatic and relevant are two things that have to be considered in assessment matters.

Most of those strategic elements are intertwined with the bell-shaped testing introduced a little while before. It is the time to deeply expose that system to the benefit of all of us, educational assessors at all levels including national educators and legislators, state educators and legislators, district levels, public schools, technical schools, college educators and university educators. It is built based on Benjamin Bloom's *Taxonomy of Educational Objectives: Book 1, Cognitive Domain*.

It is of importance to objectively present Bloom's education objectives, firstly, analyze them, secondly. After the analysis, it will be possible to come with another alternative of taxonomy of educational objectives that maybe similar to that of Robert J. Marzano's book *Designing a New Taxonomy of Educational Objectives* published in 2001. Next, we will present the structure of what the test should look like while pinpointing rationale behind it.

Brief Presentation of Bloom's Taxonomy of Educational Objectives

Benjamin Bloom presents six educational objectives in his book *Taxonomy of Educational Objectives: Cognitive Domain*, published in 1956 by Longman and renewed in 1982 by the author himself and David Krathwohl. This book contains two parts. The first part is composed of three chapters in which the author introduces six educational objectives associated with an educational program called curriculum which should be built in such a way the educational objectives be attainable. The second part of that book studies the objectives individually by demonstrating how their reach is possible.

In the first chapter of the first part of the book, Bloom (1956) declares he just brings problems to the educational system by the fact only he points his fingers at the testing system. Because testing implies previous teaching and learning, and measurement, there is some kind of relationship between measure of different objectives which are determined by the teaching content and the teaching method used to transmit knowledge and skills to the students so they can have abilities allowing them to solve problems and to be successful on the assessments. This duty has never been easy for the educators, including teachers, principals, and state assessors.

Benjamin Bloom continues to communicate his day-to-day teaching experience which would give birth to the idea of six educational objectives in his mind pinpointed and expounded today in this little book. Taxonomy, said the author, should be extremely useful, for researchers, teachers, and students.

Bloom (1956) classifies the educational objectives intelligently by mentioning the simple objectives first and the complex ones after. The simple objectives are knowledge and comprehension without which nothing can be done with the acquainting knowledge. One only can ensure that he or she knows something after he or she analyzes that thing that appears to mean something to him or her. From this point, the person is able to apply this knowledge, and to make a synthesis, and evaluate it. It is the exact time to schematize the educational objectives the way they are presented in Bloom's book (1956) to make that presentation more explicit. The educational objectives are the followings (p. 18):

1. Knowledge
2. Comprehension
3. Application

4. Analysis
5. Synthesis
6. Evaluation

Benjamin Bloom calls those six educational objectives again cognitive domains. And because they are the objectives or goals, there should be some expectation called expected outcomes represented by symbols that are, in his logic, the teachers and students' behaviors. The ultimate ones are the students' behaviors, and there is no greater behavior that should be exhibited by the students than demonstration of knowledge, skills and abilities. The greatest way students can show their understanding is by passing the state summative assessments or any assessments designed to judge their performance or determine if they should be promoted or retained.

Bloom also designs different skills and abilities appropriate to each educational objective that the students need to possess and utilize in order to be able to achieve greater scores for high-stake tests. To Bloom, ability is the result of the union of knowledge and skill. Therefore, ability and skills should be developed efficiently.

In chapter two, Bloom defines knowledge as process of remembering things as they have been encountered. This is the first way of considering knowledge. However, one may have another view of knowledge which is higher than the first one. It is the effort of human not only to recall the encountered knowledge, but also to go beyond so he or she can be able to analyze it, apply it, make a synthesis, and evaluate it.

In chapter three, he distinguishes three levels of definition in each category. Each definition is given by using a verbal description for each class and sub-class. He recommends clarity for the definition or description.

The second level of definition is given by a list of educational objective, and the third level of definition is "to make clear the behaviors appropriate to each category by illustrations" (p. 44). It means that Bloom provides test questions related to each category at different levels of knowledge, skills, and abilities.

In Part II Bloom gives details on how students can accumulate knowledge, recall what they have learned in the past, and how they can make that piece of knowledge useful. It is sure that if one gives himself or herself some time to read that part of the book, he/she will find a lot of useful information related to tests illustrating the levels of knowledge mentioned above. What is important to do now is reporting objectively the educational objective with the different levels so one can determine how many questions a test can contain.

Benjamin Bloom presents in his book twelve types of knowledge, and he contends that the student should know them, and be able to use them in convenient time and their related context. Here are the types of knowledge as they are in Bloom's text:

1. Knowledge of specific
2. Knowledge of terminology
3. Knowledge of specific facts
4. Knowledge of ways and means of dealing with specifics
5. Knowledge of convention
6. Knowledge of trends and sequences
7. Knowledge of classifications and categories
8. Knowledge of criteria
9. Knowledge of methodology
10. Knowledge of the universals and abstractions in a field
11. Knowledge of principles and generalizations
12. Knowledge of theories and structure

Benjamin Bloom classifies those classes of knowledge or knowledge in general in the category of what psychologists and cognition theorists call recall and things related to our memory. If the memory fails to retain any retrievable information, one will not be able to work anything out because the table would be empty; there will be no knowledge to think about, no way to make inference, analysis, synthesis, and evaluation.

As it is said, those are knowledge that come to human in their primary form, and they are classified from simple to complex. For example, knowledge of specific and knowledge of terminology, and specific facts are much easier to be recalled than knowledge of ways and means of dealing with specific facts because the latter is related to procedures which makes it a little complicated. Robert J. Marzano and Donna Walker Tileston call it "procedural memory." Tileston (2004) cites Jensen (1997) to contend that retrieval of this type of knowledge is dependent on how it was put in the memory. It is needless to tell how recall of knowledge of principles and generalizations, and knowledge of theories and structures are complicated.

What about, now, the second educational objective which is "comprehension"? If students comprehend a thing, it is knowledge, but not emptiness. It means that comprehension is comprehension of something that is not useful until reflection is applied to it. It needs to be explained, compared, if possible, to some other things, or combined with them in order to produce a new thing which may be, an expected result. This expected result may be the students' behavior, the result of the education objective:

comprehension without which no information, recall, or knowledge is useful.

How can one understand if he or she has no skills and abilities? Comprehension or understanding is demonstrated via skills and abilities. It is the reason why, Benjamin Bloom defines them immediately after his exhaustive expose of knowledge that is the first element that comes to the brain to be stored which will be recalled unless it is understood and which is impossible in the absence of skill. Therefore, there is a brutal knowledge that is not true knowledge unless it is comprehended, and true knowledge after epistemological reflection to be transformed in what we call episteme. Here is Bloom (1956)'s definition of skill and ability:

> Organized modes of operation and generalized techniques for dealing with materials and problems. The materials and problems may be of such a nature that little or no Specialized and technical information is required (p. 204).

Bloom (1956) distinguishes three levels of understanding (translation, interpretation, and extrapolation). Each one allows the students to deal with some kind of materials and problems. When he says that for some knowledge a student may need no specialized and technical information, it means that for some other knowledge the same student may need them dependent on the nature of the knowledge. The more knowledge is complex the more one needs specialized and technical information to deal with the situation.

For example, one needs only to understand the content of a message to translate it, to tell other people about it in his or her own words without taking the essence of that message away. However, that person needs to go beyond the author's message to explain it. In addition, he or she needs to read between the lines and question the hidden meaning of the spoken words to find out what the speaker intended by what he or she says loudly. Nevertheless, interpretation may not have this uplifted dimension if otherwise the assignment requires extrapolation level. It may be required at some college and university levels. Will students be able to perform at that level of understanding?

Anyway, understanding is only the second stage of Bloom's educational objectives. The third one is "application." One wonders that it is impossible to apply what he or she does not understand. It means that we can feel how Bloom integrates the levels of knowledge, comprehension of the knowledge imposed to one sense and spirit. We can also feel how Bloom integrates different skills and ability levels that one needs to apply when necessary based on what he or she has learned.

Just remember, for now, that assembling the three first elements is necessary, which should not be forgotten, of the educational objectives trajectory. Those are: knowledge, comprehension, and application. They are logically assembled to demonstrate how instruction is working smoothly from simple to complex.

Applying knowledge, skills to a new situation has never been an easy task. Bloom intends that one may be able to apply a simple formula to a new situation, but when it comes to applying a theory, an idea, or a principle, the person would need more abilities and have a very high level of understanding. Otherwise, it will be impossible to forecast the future and to make change happen where it is necessary. The student will not need to copy or reproduce the author's experience; he or she needs to remember the experience, remember the formula, or past experience just to understand the actual situation which may turn to be a kind of threat. Situations may be different and require different approaches.

If situations can only be different and require different approaches, it means that the person is obligated to analyze the actual situation in order to be able to compare the two situations and establish the extent of their differences. It is the reason why Bloom presents "analysis" as the fourth taxonomic element of educational objectives.

Analysis, to Bloom, is the broken down of an object in its different constitutive elements. When considering those elements apart we could have a better idea of what the whole object is about. Bloom differentiates three levels of analysis which are analysis of elements, analysis of relation, and analysis of organizational principles.

Analysis of elements appears to be the simplest analysis ever because they are simply elements. The simplest element may be understood in its simplest way because it is presented to one's view as it is and it cannot be other than what it is. Even there, one can make mistakes and misunderstand the simple object or idea for any reason, including lack of observation, questioning, insufficient time allotted to the inquiry, or simply because our own idol to borrow Bacon (1620)'s approach of digging the truth.

Fortunately, Benjamin Bloom introduces another analytical level, which is comparison. If mistake has been committed comparison of objects or ideas would give the analyst a new chance to correct the mistake and take back the track. What is more important to highlight and anchor, in this endeavor, is the fact that there is no more one element, but many of them. What could be one's discovery? The use of good method, logic, and clear wavering thought focusing on the reality or the imposed situation only will allow the student or the thinker to arrive to a better conclusion. That conclusion is may be what Bloom calls synthesis.

Before we think about Bloom's fifth level of educational objective which is the synthesis, let see the third level of Bloom's analysis which is "analysis of the organizational principles." This is a stage of systematization, arrangement of thoughts in structures and patterns. Students should be able to identify ideas, to study them systematically, understand them in their context, analyze the content of the message, different ideas, and be able to make a benchmark study so he or she can classify ideas orderly. It is only after this performance that one may be able to make a synthesis.

Synthesis is the level of creativity, level of production. The students need communication and writing skills so they can express their level of understanding, convey their feelings. In addition, the students should be able to organize their thoughts clearly. Bloom (1956) distinguishes again three levels of synthesis which are:

1. Production of a Unique Communication. It implies that students need to be able to use their own style and utilize their own words.
2. Production of a Plan, or a Proposed set of Operation. A plan is what allows the writer or speaker to produce an organized pieces of writing or excellent speech that addresses well a subject. When the student cannot do this, he or she may commit mistakes, may be not being able to stay on the subject, or pose the problem wrongly. Therefore, it is impossible for him or her to produce the expected and proposed work.
   Synthesis is what leads to derivative works.
3. Derivation of a Set of Abstract Relation. From meta-analysis, the thinker can propose a new way of understanding things by formulating hypotheses whose purpose is to "explain particular data or phenomena" (p. 207).

Bloom (1956)'s sixth taxonomy of educational objective is evaluation. It is the highest level of production based on which the producer is obligated to validate his or her production and look for its utility related to literature and the reality. This is the deepest level of epistemological method of verifying the quality of the production. In other words, Bloom is looking for evidence, and in order to reach it, one must establish criteria based on which he or she can verify the new idea as such. This level of reasoning is so high that Benjamin Bloom thinks that criteria may be given to the students by the teacher. This is the highest level of thinking and highest assessment a student would have ever been given.

Bloom's Taxonomy of Educational Objectives' Lacks

When considering Bloom's taxonomy, one can have impression that Bloom has produced the most perfect educational work, and there is no need for modification or change because it's a finished work. It is true that Bloom has rendered a wonderful service to the world of education, to the teachers, to the principals, and the state assessors, but there are some empty spaces and holes that one needs to fill up in order to make this literature perfect.

Evidences are so clear that Bloom himself declared that he is not providing solutions to the educational system, but problems instead. He deserves credits for his intellectual probity, and we should recognize that he was the herald and the educational prophet of his time when it comes to establishing educational goals and method of testing. No one can deny this as a fact because since the publication of his book, educators started thinking of standards laid in the preparation of a fit curriculum, teaching that matches that curriculum, posing the problems of what kids need to know, and evaluation of their understanding. In other words, after Bloom's passage, education has been taken in hand very seriously to foster the students' intellectual productions and accomplishments.

Marzano (2001) evaluates Bloom's Taxonomy in the following terms: influence and severe criticisms, self-criticism (p. 8): the problem of structure. Robert Marzano offers a new structure to help better understand Bloom and to contribute to the evolution and development of teaching (establishment of educational objectives and goals) and evaluation (assessment of learning for better achievement). If the Bloom taxonomy was so perfect, Marzano (2001) would not feel a need to propose a new taxonomy. He criticizes the placement of evaluation classified as the sixth educational objective. He thinks that evaluation is not necessarily the last stage of educational activity because it should help determine the value of the primary accumulative knowledge. Marzano (2001) reports Bloom's words as self-critics in the following terms:

> Although evaluation is placed last in the cognitive domain because it is regarded as requiring to some extent all the other categories of the behavior, it is not the last step in thinking, or problem solving. It is quite possible that the evaluation process will in some cases be the prelude to the acquisition of new knowledge, a new attempt at comprehension or application, or a new analysis or synthesis (pp. 185, 9).

This passage indicates that the author was aware about the mistake, but he did not regard it as such. It appears to be lack of reflection which led him to living in a state of illusion. The fact is that Benjamin Bloom forgot that he wasn't talking about instruction, acquaintance of knowledge, but about creation of programs which require evaluation at last. However, the first acquainted information in the process of learning needs to be tested and evaluated to ensure that its validity, its relevance, and appropriateness. The primary contact with the reality, the nature, the object whether it's physical or spiritual is the basis of all knowledge. Therefore, if that basic element is not what it appeared to be, and we accept it a priori, the researcher is wasting his or her time and is selling his day for free. We should have a good beginning in order to expect and have a good end. The house cannot be firm if its foundation is not. Therefore, evaluation had to come third and/or come back repetitively, or after each step of the learning process. Information or the primal or the brutal fact should not be stored for retrieval if it is not true knowledge; it could be a poison ivy and catastrophic for one's mind.

However, Marzano does not insert evaluation in his new taxonomy. We don't know yet if he has any equivalent of it in his stratifying educational objective stages, including the following:

Level 1: Retrieval
Level 2: Comprehension
Level 3: Analysis
Level 4: Knowledge/Utilization
Level 5: Metacognitive System
Level 6: Self-System (p. 30).

We are not here to comment on Marzano (2001)'s New Taxonomy of Educational Objectives, but on that of Bloom (1956) that is the most perfect and great contribution to the educational system. Benjamin Bloom just made a couple of mistakes. We do not want to be too scolding toward him. He started what we have to finish today as Robert Marzano is trying to do it. We can remark that Marzano (2001) does not mention the word evaluation at all, and he adds the word "utilization" accompanying the word "knowledge" as the forth level of his ranking taxonomy of educational objectives. It could replace Bloom's "application." Anything that can be applied to any contingent situation is pragmatically useful and relevant.

Further, Marzano (2001) remove Bloom's fifth and sixth levels (synthesis and evaluation) and replaces them by the followings educational objectives which are "metacognitive system" (level 5) and self-system (level

6). As one knows it, metacognition is reflection on cognition, on facts and ideas that impose themselves to us as knowledge. When we reflect on that piece of knowledge enough we may be able to come to a conclusion.

Metacognition is used here to make a play with words. There was no need to replace synthesis by it as metacognition does not really play the role of synthesis although Marzano (2001) contends that metacognition encompasses monitoring, evaluating, and regulating. Metacognition is more likely to be analytical reflection on them in order to make a conclusion.

Marzano (2001)'s last step is self-system which cannot replace the word evaluation. It is rather a psychological dimension based on which the system is concerned with people's attitudes, beliefs, emotion, and search for balance and equanimity. This could be related to affective aspect of Bloom's taxonomy. We mean that evaluation is considered as another dimension where the assessor establishes criteria based on which knowledge should be tested and classified for what it is exactly epistemologically. At this time one can think about the application of any types of knowledge or ideas, or theories.

Before Robert Marzano, some other educational theorists have tried to replace Bloom's taxonomy. Among them figured Dr. Norman Webb who developed his method of teaching and assessing in 1997, which he called "Depth of Knowledge" (DOK). It was a new taxonomical approach containing four levels, including recalls and reproduction (1), skills and concepts (2), short-strategy thinking (3), and extended thinking (4). Dr. Norman Webb's method was published for the first time on the web in 2009 at http://www.MDE.k12.MS.US.

Webb (2009)'s DOK does not criticize Bloom's taxonomy. Instead he tried to provide another alternative of educational objective the same way Bloom did it moving from simple level of educational objective to more complicated ones. However, we assume that Norman Webb had some negative untold critics against Bloom to be willing to develop an alternative taxonomy parallel to Bloom's taxonomy presented to the public forty-one years before.

The last taxonomy that intended to replace Bloom's taxonomy, because maybe of its lacks, is the "Structural Observation of Learning Outcomes (SOLO)" was developed by John Briggs in 1999. Briggs's objectives were to relate teachings and desired outcomes to assessment. Any educational approach that proposes to do this intend to put emphasis on the fact that Bloom did not influence enough or at all the way educational leaders build curricula (Marzano, 2001). Without doubt, we think that Briggs also found something that was not working in Bloom's taxonomy.

We feel that it is just time to provide our own classification of educational objectives using Bloom's elements. There will be no change of words, objectives, but change of places. It means that we are going to reorder Bloom's taxonomy of educational objectives for many reasons that we will tell later. It is not important to change the words because they are the exact ones that had to be used, but in their right position. Let's present the new classification first before we can tell the rationales behind them, and this is where one can recognize some of Bloom's lacks. Here is the new classification:

1. Knowledge
2. Analysis
3. Comprehension
4. Evaluation
5. Application
6. Synthesis

Rationales behind This New Classification of Educational Objectives

As it is said before, it is not necessary to give many details related to the nature of each objective and different aspects they contain based on their simplicity and complexity, but some aspects of their nature related to their position in the classification. Therefore, it is a necessity to explain why one educational objective should be placed after another one but not the other one. This is, may be, the unique way one can move safely from one level of knowledge to another and be sure that what we know is the truth, but not mental cliché (Horkheimer and Adorno in the mid-twentieth century) (Touraine, 1995) or an idol (Bacon, 1620). At this point, we will ensure that this piece of information, or the result or the outcome of the encountering of the fact or the idea and the mind, is applicable.

When Is Knowledge Applicable?

Knowledge is applicable or usable when we are sure that it is what we know it is in itself and when it can help solve our daily problems. What is knowledge now? Knowledge is information about something of any nature (material, life in general, the reality as phenomenon, or the reality related to inner thoughts or the invisible world). According to Jean-Paul Sartre, the famous French philosopher and Martin Heidegger the famous

German philosopher, it seems that it is extremely difficult to have any notion of knowledge of things or even of self. As we exist for ourselves and for others, we need others to know us as much as we want to know them, but unfortunately, each time we approach others, they flee us.

This is again unfortunate that many theorists deny all spiritual realities and confined knowledge inquiry or truth to observable things only (Skinner, 1953, cited by Phillips and Burbules, 2000, p. 9), and because of that they may make it very difficult to find the truth or certain relative truth based on warranted assertibility for truth (Dewey, 1936). They are limited; they limit themselves, and other researchers. They may not be serious because human being is not an object, but a being with psychological aspect. There may be things (thoughts for example) that are not accessible to the inquirer; therefore, thought like any other psychological fact may not be observable. James (1950) contended that thought belongs to personal consciousness, and it is the most difficult task for a researcher investigating it if the owner does not reveal it.

How can one find the truth in this situation? In addition, they have to remember that human being is not only an animal, but a reasonable animal as Aristotle and Descartes could say it; therefore, we have to explore this rational dimension based on reason and logic and deductive syllogism as method of inquiry (Aristotle and Dewey). Sigmund Freud has rendered a great service to humanity when he came with his psychoanalytic method. Truth is there; we will find it depending on the method utilized to investigate it. Method of investigation is all. We want to express our regret to Skinner who used reason (internal dimension of being) and criticized the inner being at the same time; He is guilty of conscience.

There is no such one method (Skinner behaviorism) to be used to discover the truth meaning to apprehend knowledge of something. Skinner contended that because he wanted to deny the existence of the inner being, it just does not exist; that is not true. He is wrong; he needs to prove it. Because his arguments are weak, he would not have been able to prove it as a fact. He might have been erring. Discovering the truth depends on time, nature of things, space, and method utilized to discover it. Sartre contended that the choice of the plane only can help in discovery of the identity in itself. He wrote: "It was only a matter of choosing the appropriate plane" (p. 75).

It is important to support the American pragmatists' methodological view of the reality that human beings cannot reach the truth, but they can search for it based on certain principles and make a conclusion. However, their synthetic truth should be left open for future actions which can allow to see and to observe different aspect of that piece of information (Charles

Peirce, Williams James, John Dewey, and George Mead, Rosenthal, Hausman, and Anderson, 1999; and Phillips and Burbules, 2000). Those American philosophers meet what Heraclites could call long time before "the mid-way of inquiry" mounted on the following statement: One will always have a second half to do.

One should know that knowledge is a brutal fact, and encountering thing that is always astonishing by its abrupt nature (it's where science is born). Disclosure of that brutal fact is dependent on its source, its clarity from the source, its disclosure availability, and many other circumstances. Its contingency may cover it with aspect of applicability for this period, but not another one, and for this region or country, but not another region or country. This is why we agree with John Dewey when he advises practitioners to define and redefine education based on the people's needs, their period, and country.

Because knowledge a priori is a brutal fact, one should not assume that he or she knows the truth at the encountering moment. The researcher must question the fact or the imposed idea, analyze it, and observe it for a reasonable time. After that he needs to test it and verify that it is true knowledge which is applicable in the reality in different circumstances or a particular one.

Attention: if that piece of information is used in all circumstances as universal knowledge (whether it's a theorem or axiom, or not), one can have good results while the other one may fail, or the same successful person can fail too later because of limited fruitfulness of knowledge. This viewpoint is not only good for educational leaders to know, but also political ones. This is why the postmodernists and post positivists are questioning the modernity and are eager to deconstruct its pretended secured established knowledge, rationalism, or empiricism. The father of modernity, Rene Descartes, in his methodic skepticism, said that he will never affirm that something is true before he realizes that it is so. It means that the encountering information needs to be tested and scrutinized under acerbated critiques before it can be used as applicable information.

The piece of a priori knowledge being a brutal fact is the primary knowledge that one has that may be useful or not. In case it is useful, it is a true knowledge and its applicability and usefulness are evident. Evidence is all the scientists and researchers' concern. Something that is useful, applicable, evident must be considered as a valid thing. Validity is what makes a piece of information or knowledge applicable or useful. Without it this information should be rejected as true information and knowledge, and this is the basics of all scientific inquiries.

Based on what is said above, one cannot think that it is possible to know something easily; one must understand the object that needs be thought inside out or profoundly and verify it as such before yelling victory. This is the most difficult aspect of taxonomy. Therefore, Benjamin Bloom made a mistake when he classified understanding as the second educational objective in his taxonomy of educational objective. What, in fact, had to be placed secondly? The answer seems to be analysis.

Analysis

Considering the complexity of knowledge and how it is difficult to get acquainted with things, we think that Bloom (1956) made a mistake when he placed comprehension after knowledge in taxonomy of educational objective and analysis in the fourth position. It is impossible to know something very well at a glance. We think that analysis should be classified secondly and immediately after knowledge. We cannot understand something for what it is without analyzing it. When it comes to comprehending a thing, analysis is king.

There are light analysis and profound analysis. Light analysis is done in a very short period of time when a person tries to understand simple things. He or she may operate mentally by asking and answering questions, by decomposing elements in their most simple forms. If that person has any skills or abilities to do it alone, he or she often addresses to another person. This is the most direct way people acquaint and accumulate knowledge. True knowledge is the result of inquiry and there is not inquiry without questioning.

When it comes to understanding complex things we do profound analysis. Details related to that new knowledge and judgment are necessary. Whether the analysis is simple or profound one still needs to ask question without which it is impossible to understand things

Why should one ask question about anything? The answer is just a quest about knowledge. If the primary knowledge we have of something was the truth, it would not be necessary to ask questions anymore. We do so because we are skeptic and we are not satisfied with the seeming knowledge that we have. The seeming knowledge is the phenomenon, but not the object in itself, or not its internal dimension, the being-for itself, the idea, or the inner being or the eternal being (Emmanuel Kant, Martin Heidegger, and Jean-Paul Sartre). Sartre (1943) wrote, "Knowing as for its ideal being-what-one-knows and for its original structure not-beingwhat is known" (p. 218). Knowledge is slipping away every time. What we think that we know is not knowledge. Therefore, it is necessary that one makes an epistemological analysis before he or she thinks, or agrees that he knows something. Do not pronounce the word understanding without prior analysis of knowledge.

Related to instruction, a teacher must check out very often his or her students' understanding and make sure that their "yes, I understand" is a true "I understand." The teacher must be sure that all the students understand because understanding is the center piece of all communication without which all teachings and contracts and meetings are wastes of time. The teacher can't go any further if any single student is not very sure that he or she really understands what is being taught. The ultimate goal of education is to make students understand the curriculum taught so they can progress in their learning experiences and improve in the classroom assessments, which are formative assessments in order to succeed on the state assessments, which are summative assessments.

As a method of checking for understanding, the teacher should check several times at a time, sometime by pointing fingers on particular students individually:

a. Class, do you understand?
b. Rolling eyes, do you understand?
c. Pointing finger, do you understand?
d. Are you sure, sure, and sure?
e. Asking questions based on the content curriculum being taught sometimes in group, sometimes individually. By doing so, questioning and understanding may become a habit for both teachers and students.

Grant Wiggins and Jay McTighe were highly aware of difficulty of understanding when they wrote their book titled *The Understanding by Design Guide to Creating High-Quality Units* that they asked a skeptical question about the teachers' willingness to make their students understand. They contended that it makes sense when the students have possibilities to express their understanding. They also, to express the complexity of understanding, provide six individual tasks the student should be able to do when they understand. It means that if they cannot do them, they do not understand, or they do not fully understand. What is the missing point? To answer this question, it is necessary to enumerate the six tasks of Wiggins and McTighe (2011) that students must be able to accomplish (The six Cs of understanding). Here are they:

1. The capacity to explain.

2. The capacity to interpret.
3. The capacity to apply.
4. The capacity to shift perspective.
5. The capacity to empathize.
6. The capacity to self-assess.

One can realize that the six capacities presented above look like taxonomy or nomenclature of knowledge related to task performance. They are presented from simple task to complex ones. Congratulation to Wiggins and McTighe (2011). However, they forget the simpler one, which is what Vygotsky (1978) called "describe." To Vygotsky (1978) description is related to phenomenological aspect of the thing, but not its essential aspect. In order to know if we (the students) are able to get the proximity of the essential aspect of the object, we or the student should be able, firstly, to describe that object and secondly to explain it.

Vygotsky (1978) referred to K. Lewin to talk about phenotype and genotype. The former is explained by the appearance of the object which can be only described because the object imposes its presence to senses, to the world. The latter is explained by the inner part of the object, what makes it up, its real nature that is the hidden part, the origin of the object. Does a teacher want his or her student to describe the object or explain it?

A teacher should need both the phenotype (description of the object) and the genotype (explanation of the object). However, if the student can describe the object but can't explain it, he or she is very limited and needs more knowledge. One really knows an object when only he or she can explain it. Being able to explain an object means that you have the deepest knowledge of it, or you understand it. In order to be able to do that, analysis of that object is crucial. This is why the first three educational objectives must be:

1. Knowledge (abrupt, brutal fact or information, or mind/object-encountering)
2. Analysis
3. Comprehension

Those form half part of Bloom's taxonomy of educational objectives composed of manifestation of the world to human mind. It is a period of acquaintance where the person's mind calls upon his or her brain for help so they can develop some capacity to comprehend the phenomenon and the effect of the encountering that would not leave the brain in peace for that effect will come back at any time, or very often in time of need or by simple souvenir because the brain has stored something consciously or not.

If the brain responds favorably, some efforts will be deployed to understand the new phenomenon. At this time one can describe and explain the phenomenon and explain what is behind it. Therefore, it is very possible that skills and abilities that come out of this intellectual effort will allow to apply the new understood knowledge. But, before the application of the knowledge, an evaluation of the understanding is necessary. It is, may be, what Marzano could name metacognitive approach.

Evaluation

It is not necessary to define rigorously the word evaluation here because we just want to explain why it should be placed in the fourth position. Evaluation is a way of verifying validity, truthfulness, worthiness of something. It means that evaluation is what ensures that the learning is good, and therefore, the new knowledge can be used.

How can one apply a new knowledge, disseminate information that is not verified and evaluated? The method is that the learner writes down criteria based on which he or she can accept and declare that the new knowledge is good and worth applying. So evaluation should come after comprehension and before application. It is necessary to analyze the information or knowledge to comprehend it; the conclusion cannot be applied without being evaluated.

Application

This is the implementation level where the new knowledge is fundamentally what we know it is, but not the way they say it is or it appears to be. Application is the stage of action, change, and transformation. When students learn seriously, understand the curriculum taught, it is made easy for them to improve and to have a greater achievement. Analysis allows the learner to have skills and abilities to accomplish things and solve the most difficult problems. This is at this stage that one can work out tasks pertaining to explanation, interpretation, application, shifting perspective, empathizing, and self-assessing, according to Wiggins and McTighe (2011).

What if students have the greatest understanding, the necessary skills and abilities to perform a task but they still can't perform it anyway? Four things can be the causes: fear of exams, memory problems, not knowing how to take exams which will not be developed here but left to the care of the teachers, and lack of exam setting structure. Those serve as barriers to the application of knowledge, skills and abilities that the student possesses to perform freely, limitlessly, and intelligently. They need to be apprehended individually to help remove obstacles to high performance. Is it the right time to present three of them being extremely crucial?

a. **Fear of Exams:** fear is one of the greatest barriers that make students in situation where they cannot take exam peacefully. They knew how to respond to some particular test questions but they are confused at the moment of the test; they do not have clear thought anymore and consequently they fail to answer right the questions. Just after the exam, they become weird again because they were ready for any type of exam. There should be a problem. Did they have previous bad experiences with taking exams and they are traumatized? Is the environment a problem and people are hostile? Or do they have lack of confidence? Those could be different factors to create fears in the candidate to be shaken up with chuckled body parts.

Sometimes, they do have confusion, and they forget everything. Any mental operation needs support, something to lean on, to think of and to analyze. On the contrary, nothing can be done. It is like math and physics problems; if you forget the formula, you will be in difficulty to solve related problems. It happened to me when I tried to solve an optics problem on my baccalaureate part I in Haiti. That was a very serious exam that one cannot afford to fail. The results of that exam was published via radio, and it creates a lot of emotions at that moment you sit next to your little radio accompanied by family and friends. When they call your name you feel a lot of joy and your family and friends jump all around. However, if your name is not called, you feel extreme shame that you think that life is over. It happened to me because I could not solve a physics problem, and the cause was that I could not remember a formula. I was so confident with doing optics problem, I held that I had to do it, and I did not switch to my other option which was electricity. Memory is a big deal in learning.

b. **Memory Matters:** Without it learning is impossible. Memory is the ability that one has to store data in his or her brain and to retrieve or recall it in time of need. There are different types of memory, and it seems that information is also stored in different parts of the brain. The Information is saved in a part of the brain but not somewhere else based on the nature of that information and the objective of the storage. There are semantic memory, procedural memory, short-term memory, and long-term memory. According to Tileston (2001), semantic memory is mostly used in the schooling process.

Where is the problem of memory now? Where is it located at? The student starts having problem when he or she stores information in very short period of time and wants to retrieve it. It is better to slowly store the data and to retrieve it faster later in time of need (Tileston, 2001).

What is semantic memory? According to Tileston (2001), semantic memory is the type of memory, located in the cerebral cortex, mostly used in education. The information saved there is retrieved only if rehearsal has taken place. It is also based on the saving procedures utilized. If information is stored intelligently and it is moved from short-term memory to long-term memory based on what psychologists called working memory, it can remain there for life. In this, teachers can help.

Procedural memory is easier to be used because it is practical like learning how to ride a bicycle or how to cook. However, it is not so easy to remember the entire process unless rehearsal is practiced by the individual.

Besides memory problems, which can be the major cause of failing a test, students may fail the exams because only they do not know how to take a test. And the last cause of failure is the structure of the test that is lacking.

c. **The Structure of the Test Matters:** Lack of structure of the text exams is the main cause of students' failure. This cause of failure is exogenous to the students. It means that they are not the cause of their own failure, but the assessors who fail to create tests that go from easy questions to hard ones. Easy question refers to memory or recall questions that do not demand too much concentration for one part. For the other part, hard question refers to questions that require much more attention and concentration. It is more likely for students to answer well memory questions than the inferential ones.

As there are many types of memory, students would have answered well many questions to help them score higher. By the same token there are many kinds of inferences arranged from lighter inference to higher inference. There are 70–80 percent of chance for them to answer well inferential questions. As the structure of the questions forms a cycle, the same types of questions go and come back until the end of the test and the students' mind is refreshed and renewed throughout.

It seems that we are making progress at this point with two other educational objectives which are "evaluation and application." Remember application comes at the end or before synthesis; evaluation comes in the fourth position and can be repeated after each stage because one must ensure that he or she is progressing in the right direction. It should be done so to make the last check even though it is the most systematic one. One thousand times on your art review your work; erase often, but seldom add (Nicolas Boileau-Despreaux, theoretician of seventeenth century's French literature). Therefore, it is good to repeat the new educational objectives here to focus on the new classification forever:

1. Knowledge (abrupt, mind/object-encountering)
2. Analysis
3. Comprehension
4. Evaluation
5. Application
6. Synthesis

All tests at all levels including public schools (Elementary, Middle, and High Schools), college and university levels, and technical schools must follow this classification of educational objectives if educators and instructors need to see improvement in their students' test scores.

CHAPTER V

Introducing Bell-Shaped Testing

Defining Bell-Shaped Testing

Bell-shaped testing (BST) is a testing strategy based on which the assessor integrates memory and skill questions related to the taxonomy of educational objectives presented by Benjamin Bloom and revised by Dr. Fleurmons. Bloom's taxonomy of educational objectives includes knowledge, comprehension, application, analysis, synthesis, and evaluation. Fleurmons has revised this taxonomy and made an arrangement by reordering them as what follows: knowledge, analysis, understanding, evaluation, application, and synthesis. BST will contain questions according to the latter taxonomy. The questions should be placed from the base of the curve to its top, and around the curve to form a structural cycle. Lastly, questions need to be placed progressively from simple to complex ones.

Goal of Bell-Shaped Testing

Bell-shaped testing's goal is to help the students have greater achievements in school and on the state tests. This is what everybody wants to happen including parents, teachers, principals and/or managers, directors of programs, and the students themselves. Instead they encounter the opposite. Why this simple project cannot be accomplished? The exams are not structured, do not follow any principles related to what has been taught by the teachers implementing designed curricula. Sometimes, teachers do not even teach based on a given curriculum; they teach based on a given book and it affects the testing system. In order for kids to achieve higher grades or scores there must be a balanced or aligned curriculum, strict implementation of that curriculum, and a testing system following what we call bell-shaped testing, which is the ultimate matter and concern here.

Objective of Bell-Shaped Testing

BST has four major objectives encompassing psychological aspect, proportion distributive aspect, content aspect of the assessment, and measurement aspect. Let's see them individually briefly.

a. **Psychological Aspect**: When questionnaires begin with easy questions for which answers are reachable by a simple recall, students have feelings that the test is not difficult at all; it's like they can breathe and have hope. Consequently, their determination to take the test increases; their mind is not shaken and trembled; instead, they are determined to answer all the questions whether they are mnemonic or not, complex or not; it is reinforcement of determination.

One should not put in his or her head that they give a gift to the students by the fact they start with the easiest questions; it is just meeting Bloom's first educational objectives, which is knowledge processed by recall. Why most state tests omit those types of questions? Assessors omit them or insert them maybe late in the process of assessment just to trick kids, to make them feel that they do not meet qualification for their levels of study they are in. A test should not be tricky; its objective is to meet the quality of teaching, reinforce students' learning, and verify students' level of understanding.

There are twelve elements that should be checked in the first level of Bloom's educational objectives. It starts with simple knowledge or recall; they should be mentioned one after another. If a student answers the first twelve questions well, he or she is more likely to score higher. After answering twelve questions very well, that student may have peace of mind, and he or she will not be afraid of spending some limited time to reflect on questions that appear to be difficult.

Nevertheless, a teacher or an assessor does not need to ask all twelve recall questions; it is dependent upon the taught curriculum for the season, except if the test is the exit test or summative test.

As state tests are always long it is possible to ask all twelve recall questions. Go back to previous page to review the recall question or questions related to simple knowledge. No matter what happens, a good educator and assessor ask questions related to knowledge first before proceeding to questions related to understanding in order to help psychologically the students.

As the new taxonomy switches comprehension position from second level to third level by replacing it by analysis, the second group of questions should be then analytical questions (3) followed by three comprehensive questions. Our math allows us to count a number of eighteen questions for this time. It means that the assessment becomes complicated because the last questions reduce a student's chance to score so higher than one could imagine. However, there are more recall questions than analysis and comprehensive questions. Now it is time to further move to the second objective of BST.

b. **Proportional Distributive aspect**: BST would like to see that students are tested fairly based on the number of questions that should be asked in each section disposed by Bloom's taxonomy of educational objectives. A fair distribution will help assure not only that the number of questions asked fairly but also that they are arranged in their respective position in the bell-shaped system to help psychologically the students being assessed. A fair distribution provides the following arrangement including:

1. Twelve recall questions
2. Three analytical questions
3. Three comprehensive questions
4. Two evaluative questions
5. One or more application questions, and 6. Three or less synthetic questions.

c. **Content Aspect of BST**: This is the third objective of BST. There would be a total of twenty-four (24) questions on a regular assessment. They can be repeated dependent on the teacher or assessor at any level plus two extra questions added to make it fifty questions as fifty is the designated number of questions on a regular tough test, Extra questions can be of any types still in the given BST structure. I suggest that they be knowledge or memory questions to increase the student's passing grade.

Content aspect of BST (an advocate trial designed to help educators in their process of testing kids) includes fair distributive system and curricular content. Fair distribution would appear to be an effort of eliminating bias and favoritism. Although the distribution is fair, it will not make any difference if the questions are not selected based on the taught curriculum. We cannot ask the students questions on the moon if the study of moon is not included in the curriculum. It would be a killing process of sending them to space eliminating their chance to pass the test. Therefore the content aspect of BST is extremely critical if we want the students to achieve higher scores on the assessment.

Bell-shaped testing can be a wonderful tool to the criterionreferenced measurement called again criterion-referenced testing invented and promoted by Robert Glaser in 1960s and James Popham's today's standard setting method. This theory had birth under the pens of other theorists before him such as Ancoff, Beuk, Auftee, etc. However, the former had preferably set standard passing scores for technical and practical professions so practices could be made for only a small group of qualified professionals. Standardized set score is not our interest for this time. We prefer to consider only Popham's standardized setting method and Glaser's criterion-reference measurement because they rather emphasized curriculum content related to the content of courses that the teachers present daily to their students.

They truly are our concerns because those courses should be extracted from the established curriculum which addresses objectives of education in general and the students' needs in particular. Glaser and Popham connected assessments with the curriculum used by the teachers.

The unique condition to know if the curriculum is good is the assessment administered to the students and analysis of their scores based on certain pre-established criteria. One must consider that this assessment should not be a surprise for them. It means that students have right to know the objectives of the courses and the teachers' expectations because assessment should be based on the taught curriculum and other considerations such arrangement for the test with them.

In addition, teachers should ensure that all kids understand what has been taught as it was mentioned before. This is all bellshaped testing claims from the system. Bell-shaped testing seems to be, based on this approach, a package; it shows how to conciliate or reconcile curriculum and taught curriculum so no element and no child, on the technicality, can be left behind. This is how mastery learning can be practical and make individual intervention effective.

d. **Measurement Aspect**: Since the apparition of "No Child Left Behind" of President George W. Bush, teachers feel pressure to assess more often the students to measure the quality of their teaching and the students' learning experiences. They conduct formative assessment to prepare and make kids ready for the state exams. In fact, since 2005 and 2006 it has been a requirement for all states to conduct assessment. Test results would be indicator to help know the level of accountability of the students-scores based and teachers-quality based. However, the results are not so good because most teachers are complaining about likeliness of nonexistence of "today's required educational test" (Popham, 2003, p. VI). This legislative aspect and practical technicality can find their foundation in the bell-shaped testing system. Bell-shaped testing can help balance teaching and testing content value and time related. Popham (2003) contended that test designed should be the determinant of quality and quantity of times that should be allotted to the teaching activities, which can be effective under the utilization of the bell-shaped testing.

It is going to be necessary to think about how to arrange the test questions in the bell shape. This idea makes us see immediately the sides of the curve, its top and tails. But before we do so, we would like to ponder some innovative ideas related to standardized test, to students' scores that may have to see with individual differences and how to eliminate gaps by a proper educational distribution. Let's call this part corollary.

Corollaries

Do Not Forget What you Are Measuring

The first thing educators should never forget is the content of their teaching and what the content of their testing questions should be. Educators make a lot of mistakes when they think of measurement and propose to measure without any precise object, without measurement goals and objectives. Consequently, they will never get satisfactory outcomes. The problem is that they do not teach based on the test designed to measure their students' knowledge and skills. Knowledge and skills that should be taught should be found in the curriculum and ought to be the test content on which students must be informed as a matter of fact. Unfortunately, most educators miss these points that are necessary to make the students successful. Elliott, Kettler, Beddow, and Kurz (2012) criticized this aspect of testing in the following terms:

> Most of us can remember a testing experience, whether for low or high stakes, whether the test questions covered content that we had not been taught. Many of us also have had testing experiences where the test items seemed tricky and poorly written, thus making it difficult to show what we had learned about the content the test was intended to measure.

In addition to necessity of focusing on the content of the test, educators ought to not forget who they are testing to avoid messing with themselves. Therefore there is no reason to trick the students with unclear and guess questions. Our clear warning is found in following point.

Do Not Forget Who You Are Testing

The second thing educators must never forget is who they are testing when they are preparing their test questions. Most of them think that they are testing their students, but in reality, they are mostly testing themselves. If the students fail the test, it implies that teaching was not excellent. Therefore teachers are considered as eternal failures. They should make it in such that the students understand their dispensatory knowledge in order to be able to pass the test at, at least, 80 percent and above. If 80 percent of the total class passes that test, it means that teachers earn 80 percent of teaching qualification. If less than 80 percent passes, then they fail to teach based on standards teaching requirement; consequently, the course should be taught over to those who score under 80 percent.

Bruner (1996) contended that teaching and learning are based on interaction between teachers and students and that the formers act with intent of passing on knowledge and the latters are those who are receiving that knowledge. It is a necessity for students to learn efficiently information given by the teachers. But in order for this to happen, it is of interest that teachers are themselves efficient and competent using standard teaching strategies. In order to be sure of it, they need to check if their teaching is well received by the students and at what extent they are transformed. Formative assessment will help in this endeavor; yes, it is important to test them.

This is why Bruner made a claim in his book *Beyond the Information Given* that three things should happen in the teaching process, including the acquisition of knowledge, transformation, and evaluation. The test result only is able to indicate that the first two elements have happened. Bruner is right, and we agree at a high degree with him. Teachers make grave mistakes when they think that they are testing only the students. They are extremely wrong because they are unconsciously testing themselves while testing their students.

Necessity to Address the Problem of Individual Differences Sincerely

As we have already discovered what makes some students more intelligent than some others, which includes differentiated environments (rich, medium, or poor), interaction of human with the environment, intervention of parents and teachers, other educators such as school administrators and principals, foods intake affecting brain and mind, sleep affecting brain and mind, and self-motivation, we should accept it as a fact that there is superior IQ with experiences and interaction with the environment. Let's proclaim here and now the equality and uniformity of human's mind only a fortiori and before transient learning. Therefore, there are no gifted students; all the students are equal; they just need great and fostered environment.

We mean that uniformity as such consists in effort of the teachers to equalize the intelligence of the class by eliminating the existing and observed gaps. All students of the same class are supposed to be able to succeed at the same level as Carroll claimed that the students need to be educated "in a uniform way to a common standards of excellence" (Anderson, 1985, p. 117). Is it worth doing and possible?

Yes, it is worth doing and preserving, and it is possible to do so. However, it is not the way Benjamin Bloom has proposed it. Bloom proposed a plan that requires teachers to teach over the failing kids individually based on their needs if they request it. This plan represents a way of keeping those kids away and astray intending to perpetrate the actual discriminatory schooling system in order to silently slay them. How will failing kids ask for extra teaching or curriculum by themselves if there is no guidance of any sort? Kids do not know what is good for them; they just need explanation on their situation so they know the state of retardation situation they are in. The educational system should impose the alternative remediation program policy-based to alleviate their horrible and unacceptable situation needing immediate specialist intervention.

By doing so, the system would eliminate practically what people call privilege given to a group of students to get advantage-individual differences based when it comes to learning based on differentiating intelligence which is rather on social classes and income. Carroll proved that he understood this fact when he revealed this in his text (Perspectives on School Learning/chapter, Learning Ability and the Superior Student):

> All systems of education are based on certain implicit or explicit theories of individual difference in educability. Even when they proclaim that educational opportunities are made equal for all, there is an implied premise that individuals differ in their ability to take advantage of the opportunities. (They do not say that educational achievements are made equal for all). (p. 117)

Based on these ideas, the educational systems try to support the fact that some students are more intelligent than some other students and consequently, they deserve some educational privileges. The worst is that the systems do not intend to galvanize and equalize the quality of education provided to all the students in order to fill the educational gaps so all kids could get equal attention from the educators and teachers.

Carroll refuted the British theory called "theory of individual differences" developed by the British Sr. Cyril M. Norwood in 1943 that used to be a theoretical basis of unequaled distributive education given to the students. Those educational systems, based on the Norwood Report, there are three groups of students who have different levels of intelligence and some groups are more intelligent than some other group when it comes to language and practice. This theory is falsified by Carroll who thought that there are no such types of intelligence, but a "g" or general intelligence. A student may appear to be more intelligent in one specialized domain, but not in all the domains, and they may practice more than other students for a very long time.

It means that it would have been better to create a homogeneous group of students and provide the same amount of time and the same opportunities to them before they can be assessed in the first time. In the second time, it would be necessary to reteach students with lower grade (-80, B-, or C) in order to remove any gaps among the students' ability. Then thirdly, all the student could be assessed together again. If any other difference appears, then look for intelligence or mental abnormality if we should be scientific.

One should put in mind that individual differences exist because only we have been made with different temperaments and genes, but when it comes to intelligence, we are made equal at birth, and we have received all the intelligences described by Gardner as birthrights in his books:

Intelligence Reframed and Five Minds for the Future. Howard Gardner belongs with Rene Descartes and John B. Carroll because he was not afraid to declare this paradoxical truth: there are no individual differences referring to intelligence. Differences indicated by the test scores are based on orientation, environment, choice, motivation, and practice.

The Misuse of the Word *Gift* in the Phrase "Gifted Students"

The school system here in the United States creates four categories of students which include regular students at elementary school level, gifted students who receive special education in special classes so they can learn more and faster unlike the regular students, elementary and secondary students (ESE), a group of students who receive special education based on their mental deficiency, and professional student who receive career-based instruction. It is good for the country to be able to create curricula which can promise bright future to the students. It is time to recognize that all students are gifted when considering that we have received all the intelligences as birthright (Gardner, 1995, 1999, 2006). If we create a group of students that we call gifted students, it means that we have a misconception of spiritual gifts and brain and mind development.

According to Gardner (1995), all children do not make the same experiences with informal and formal learning and these experiences need also to see with how they carry out their knowledge which is at the basis of individual differences. Here's what Gardner wrote personally in his book *The Unschooled Mind: How Children Think and How Schools Should Teach*:

> I have posited that all human beings are capable of at least seven different ways of knowing the world-ways that I have elsewhere labeled the seven human intelligences. According to this analysis we are all able to know the world through language, logical mathematical analysis, spatial representation, musical thinking, the use of the body to solve problems or to make things, an understanding of other individuals, and an understanding of ourselves (p. 12).

There would be a necessity here for neuropsychologists to start investigating new born babies to measure their unschooled knowledge of the world, and it is strongly advised after they enter a school facility utilizing a psychometrics approach. It would also be necessary to provide a uniform environment to the new born babies and make follow-up with progressive checkups for brain development and how it makes healthy and normal connection with the sensory organs. It also would be good to progressively measure their intelligence. The results of those observations and intelligences mentioned above by Gardner would scientifically indicate if individual differences really exists and at what extent.

These are not in the scope of our research. Therefore, we invite specialists and scientists in that domain to conduct related researches. We hope that it will be paid off. We prefer to go back to our regular research to see how tests in the classrooms and at state level should be conducted based on given curricula, a variety of methods of teaching that need to see with the content and structure of the tests facilitating students' greater achievements.

Structure of the test: Left and Right Sides of the Curve

Structure is a form of arrangement of items that give shape to something. There can be different structures based on the shape that the shaper wants to give to the object or to the test as in this case which is the BST. Recall questions can be placed at the base of the curve following analysis questions going up the hill and so on until the reach of the top of the curve, or its other base.

In this case, we may arrange twenty-four questions from the base of the left side of the curve to the base of its right side. All six educational objectives can be displayed on the entire curve. In case there are forty-eight or fifty questions; twenty-four items can be arranged on the left side and twenty-four on the right side and two extra recall questions at the bases (left and right), or tough questions at the near-top of the curve.

It is not enough to determine the number of questions that should be on a test; the nature of the questions is of importance. The nature of the questions should be displayed in the teaching content revealed in the curriculum. If any questions in the test are not related to the teaching content and the curriculum, then that test is invalid and is not a standardized one. That's all it is about here. Teachers need to be fair and be professional; they are not supposed to be tricky and striking. Otherwise, they would be hurting themselves unconsciously. Therefore, they need to wake up.

Chapter VI

Teaching Content and Teaching Strategy that Match the Testing Content

It is crucial to match teaching content (TC) and teaching strategy (TS) with testing content (TC). It is a must in order to keep the students abreast with what they need to know, about the content of the tests, and what is expected of them so they can do well on the tests. As we have already presented the organization of the assessments and the structure of the tests, it is important now to talk about the TC and TS to see how they can match testing content.

Teaching Content

Teaching content refers to curriculum used in the process of educating the students which should be chosen or developed meticulously. Why should it be chosen meticulously? It is necessary to respond to the students' needs, to take account of the community's needs, and to match the teaching content with the testing content. In this case, Lunenburg and Ornstein (2004) declared that "curriculum can be defined broadly as dealing with the experience of the learner" (p. 478). Taba (1962) said that the objective of curriculum is "to serve the needs of the students and to promote active learning" (p. 407), and Erickson (2002) said the following: "Curriculum is what is attached to the lives and cultures of the learners and to the world beyond the classroom door" (p. Viii).

Curriculum is related to what H. Lynn Erickson called educational standards. This approach is valid for testing. Testing cannot be standardized if it is not connected to the curriculum taught and vice versa. Curriculum is defined as a group of concepts to be transformed in knowledge to be taught addressing the students' social, cultural, and intellectual needs for their complete development. Concepts will have different forms and aspects regarding the subject matter that constitutes a corps of knowledge critical to development.

The concepts should be viewed as basic knowledge in the first time and fundamental knowledge in the second time or after. Bloom thinks that basic knowledge should be first taught before teaching fundamental and complex notions. Basic knowledge is like information that needs to be filtered, in Skinner's language, before it can be selected as what can be taught and what the students need and should know, but nothing else. It means that it is a waste of time to teach kids things that cannot respond to their immediate needs physically, socially, culturally, intellectually, and scientifically. A curriculum should be able to address problems that seem to be threats for a given society, able to help resolve conflicts, and build things to shape straight a community. Curriculum deems to be vital and practical.

In order to step further, building that important piece of curriculum is critical, and it is also crucial to select those who should work on it as group. That should be knowledgeable when it comes to development in general and curriculum development particularly. If the curriculum is not teachable, is not vital, how can we build a great society? What will we teach kids and how can we address their intellectual, mental, and moral development?

Having developed a great curriculum is one thing, but teaching it is another thing, which is the most important one. Building the curriculum is a step leading to the accomplishment of the educational objectives, but if we do not teach it, it is like we throw away the whole project. It is nonsense to write a book and not publish it; it would be just ideas that are not a book and a useful tool, but some personal writings that do not contribute to the intellectual development of one's society. They may not be worth preserving, but thing to throw away with leftover tray. A great curriculum that is not taught to the children deserves similar treatment.

Teaching Strategy

In order to use the curriculum at its greater extent, teachers should have a teaching strategy (TS). TS refers to everything a teacher does or activities he or she undertakes to transfer information, knowledge to the students and to ensure that they understand the lessons, and that they can improve and pass the tests. TS is extremely important that without it, a teacher will never accomplish his desired outcome. Choice of one of those educational (Bloom, 1971; Marzano, 2009; Tileston, 2008; and Hunter, 1994, Carroll, 1963; and Engelmann, 1985) would be necessary. Because it's so important we reserve a special corner in this research for most of them. Otherwise this research and reflection on assessment would be nullified.

Marzano (2003) recognized and expressed the importance of teaching when he wrote what follows: "It is perhaps self-evident that more effective teachers use more effective instructional strategies. It is probably also true that effective teachers have more than one instructional strategies at their disposal" (p. 78). In his books on teaching, he presented a variety of teaching strategies in fact. He presented Bert Creemers's strategy, as example. Creemers's strategy containing nine steps was presented to the public in 1994. The nine steps are:

1. Advance Organizers
2. Evaluation
3. Feedback
4. Corrective Instruction
5. Mastery Learning
6. Ability Grouping
7. Homework
8. Clarity of Presentation
9. Questioning (Marzano, 2003, p. 79)

It is not important to give details on those strategic elements; we just mention them here as support to what we said about Marzano. But it's still important to learn them and use them based on need; they may help in any given circumstance in one's teaching career. In order to use them, a teacher needs to systematize them and rearrange them to create sense and logic.

The most important thing here is our teaching strategy which may help a lot. We will spend some limited time developing them. We are afraid that they would not take all our time as subject of an entire book. However, we may provide some brief explanations to find the foundation of that useful teaching strategy. It contains the following nine steps:

10. Choosing a Teaching Method,
11. Teaching with education Goal in your Mind in general and the goal of each course and detailed objectives should thoroughly be present,
12. **Teaching Based on the Designated Curriculum,**
13. **Teaching to Help Students with Storing and Recall,**
14. **Teaching for Comprehension or Understanding,**
15. **Using Regularly Feedback to Check for Understanding,**
16. **16. Assessing the Students Based on the Teaching Content,**
17. Using Cooperative learning, and
18. Assessing Regularly the Students Using a Formative Assessment Method.

1. Choosing a Teaching Method

We do not want anybody to be confused when it comes to distinguishing differences between strategy and method. Strategy is the whole technique utilized to accomplish anything which includes a variety of steps of which choice of types of activities to undertake, and a method that creates a way of materializing the project. For example, a business that desires to increase its sale opportunities designs a credit strategy system to allow customers to pay with credit cards when they have no cash. Therefore, it increases their buying power. Strategy is like aggressive method of planning an event toward achievement of a goal in using workable methods. It is exactly what we want to happen in education and teaching.

Teachers should not start any teaching program without having in mind a teaching strategy, which will lead him or her to a teaching method. It's not different for a traveler to not move any step forward without a route map or a global positioning system in hand if he or she does not have an idea of the desired place. The Newbury House Dictionary of American English defines strategy as planning in order to achieve a goal. Inside of strategic planning one can find the method.

According to Juran (1992), strategic planning implies methodology and skills needed to accomplish a project. Planning strategy is what a lot of school managers and teachers often miss which causes them to fail?

Method is what helps set up the plan and have insight of essential elements leading to success and the truth. It gives senses to planning activities and the way of attaining the dreamed and expected goal. What is finally more interesting? Lastly, method is what takes the researcher where he or she wants to go and know whenever he or she gets there. This is true.

Method is just a shining light whose objective is pushing away evil darkness being a threat to researchers searching for the truth as Francis Bacon could say it in his "New Organon" in using his specific word "idol." He stipulated that idol constitutes a barrier hindering the process of discovering the truth, which deals with light, but not darkness. If scientifically and philosophically we want to deal with light, then we need to, in the first place, establish a true method to project light on our path as method is intertwined with light. Otherwise we will be lost in the red sea's stream. Consequently, there will be no more business for us. Therefore, one cannot depart without a method, a guiding light in order not to be at stakes.

John Dewey has always contended that a researcher may arrive to a true conclusion dependent upon his or her method of choice. He used a very practical method called "pragmatism" to stick with the reality and nourish an ever ending inquiry to discover the truth and to reexamine the finding which may be true for a period, a generation, but not for another one because of possible changes. There is no eternal truth; every generation, in every single context or country, should redefine the meaning of life and education. This way, we can adapt ourselves with the time, change and circumstances. Dewey (1938) invited researchers to look for warranted assertibility for truth as other theorists could say validity, reliability, and truthfulness being guarantee of possibility of transferring knowledge to students and to the next generations, which could resist to the test of time. John Dewey truly belongs with Heraclites, Karl Popper, and William James, "who denied absolute truth in an ever-changing universe" (James, 1991, p. Viii Preface part: Pragmatism; Phillips and Burbules, 2000).

We share with Dewey that pragmatic view, but pragmatism would be a corollary method to our principal one developed in the preceding nine steps as one should choose a method that can be related to the nature of his or her subject matter. We rather would choose Bloom's mastery learning method, or Hunter's mastery teaching which targets the students' understanding or comprehension and which allows to check for it. See Bloom, Hastings, and Madaus, 1971; Rosenthal, Hausman, and Anderson, 1999; Hunter, 2004. At this time, we hold that it is not necessary to introduce Engelmann's direct instruction theory because it is too complicated and it requires a lot of times to understand it and implement it. We would express the same impression when it comes to considering Gagne' (1988)'s instructional theory. However, when considering him, instructional theorists refer to his nine instructional systematized points (p. 164). The most important thing to know right now is the idea of method in mind since the beginning so we do not depart without it.

The greater benefit of having a teaching method in mind since the beginning is that you will not miss the opportunity to nourish the basics which is education goal in general and particular goals for each course so students can get insights of what they should understand or comprehend nourishing their sight.

We would like to present two excellent teaching methods as models to our teachers who choose teaching as career and who would like to utilize a teaching method. They are Madeline Hunter's mastery teaching and Benjamin Bloom's mastery learning. Thomas Guskey, author of *Implementing Mastery Learning* reports that Bloom's method helps a lot the teachers in their teaching and the students in their learning activities. Would it be better off that we present only the methods' plan without details?

a. **Outline of Madeline Hunter's Mastery Teaching Presented by Robin Hunter**
   - **Objectives**: including choice of methods, principles guiding students' past experiences and providing example for principles and content of current learning, and design of teaching that helps recall.
   - **Anticipatory Set**: reflection on material presented and on the importance of learning.
   - **Input and Modeling**: teaching phase, instruction using the material and content, and presentation of new information.
   - **Checking for Understanding**: four methods of making material meaningful.
   - **Guided Practice**: providing to students guidance for practice and supervising them.
   - **Independent Practice**: students are given the way of practicing the skills learned without supervision.
   - **Closure**: last thoughts on teaching, principles, methods, understanding making sure students will take away with them a useful learning package for change and achievement.

b. **Outline of Benjamin Bloom's Mastery Learning in Twelve Steps**
   - **Objectives**
   - **Formative Assessment not for Grade**: to find out what students have already know and come with to the class as backgrounds.
   - **Correctives**

- **Formation of Small Group (two to three members) Study:** As each student is unique and uniqueness makes them have different levels of understanding, they cannot produce at the same pace. However 90 percent of them may be able to do it if they are given a certain amount of time. Therefore, they need to be evaluated individually in order to receive one on one tutoring.
- **Tutorial Help in Case of Need**
- **Presentation of Textbooks or Materials**: students are given special or recommended texts.
- **Workbooks and Programmed Instruction Units**: for special students who have difficulty understanding the procedure in the textbook form. It implies that specific tasks are given to those particular students.
- **Audiovisual Methods and Academic Game**: use of filmstrips and short motion pictures for individual students in needs. Academic puzzles can be used too.
- **Alternative Learning Resources**: not too clear because of just repetition of previous procedures.
- **Outcomes**
    1. **Cognitive Outcomes of a Mastery Learning**
    2. **Affective Consequence of Mastery**
- **Second or Last Formative Assessment.** Formative assessments are not graded or counted for passing grade. If more than one exam should be graded, then teachers may do more than one summative assessment during the semester/period/year and allow each of them a percentage of grade which could be accumulated to form the final grade.
- **Summative Assessment**

   Reference: Bloom, Hastings, and Madaus (1971), Handbook on formative and Summative evaluation of student learning. McGraw-Hill. USA. pp. 48–56).

   Notice: Mastery learning was developed for the first time by Carleton Washburne (1922), Henry C. Morrison (1926) whose method was very popular in 1930s, but it couldn't be expanded because of lack of technology (Peter W. Airasian, Benjamin S. Bloom, and John B. Carroll in Mastery Learning published in 1971, edited by James H. Block). Washburne and Morrison, in the early 30s, had extremely influenced Bloom who wrote *Taxonomy of Educational Objectives* in 1956 and *Mastery Learning* in 1968 and 1978 parallel with Freed Keller after both of them have also read John B. Carroll's book "*A Model of School Learning*" published in 1963 (Gentile and Lalley, 2003). We would express the same impression when it comes to considering Gagne's instructional theory.

Carroll (1963) was supported by Bruner in 1966, Glaser in 1968 (Bloom 1971, p. 45). Therefore, Bloom was not the first theorist who developed the mastery learning theory as he himself declared it; he just received seemingly all the credits for being recognized as the founder of this theory. Let's just recognize that he expanded it and led it to a point where the public could use it, but he was not its father. It seems that Bloom represents the peak of this theoretical history in this domain with his ideas of small group learning, individual evaluation (formative and summative assessment), feedback and individual corrective action as needed, and enrichment (Guskey, 1997).

However, it is good to remark that Benjamin Bloom's presentation has a lack of systematization because the elements were not classified as one could see them in the presentation above. However, Bloom buffered Carroll (1963), who first developed the idea of mastery learning process. Bloom just implemented and developed at a larger extent this learning theory. Therefore, he has been recognized by a lot of schools that use him. He received full credit for that cheerful and beneficiary salutatory theory. We invite those schools to refer to our systematized presentation of Bloom's mastery learning for a greater implementation.

2. Teaching with the Educational Goal and the Course Goal in Mind

Goal is the essence and basic motive of all activities; without it in hands, undertakers are wasting their time. It would have been better to stay home instead of going to the sea and throw stones while losing their precious thrones. Undertakers should aim at something, any accomplishments they could envision to reach at a point of time.

One can be positive about a person who would never undertake an activity without a goal as it is also certain that that person may have wrong goal because of not knowing exactly what he or she is doing, or not requesting salutary advice. Those thoughts are applied to teachers at all levels. Teachers should never miss any occasion to teach their students based on the desired educational goal in general aiming at change, transformation and aiming at students' understanding of the actual course and the projected ones yet to come.

Expression of a particular goal and its Importance

Each course should have a particular goal and objectives to be taught competency-based. This is why a teacher who takes engagement to teach a particular course should know why he or she is teaching it, and know the subject matter, the concepts that should be exposed in order to be able to make the students understand it. Not only should he or she be able to do it, but also be able to clearly state the goal (s) and objective (s) of that particular course so the students can know what is expected of them. This is where students' success is originated.

The goal is the reason why the teacher is teaching the course, and the objectives are the means that will be utilized to reach the goal because they represent different aspects of that goal to be taken down one-by-one for a better and sure reach. If applied, teacher should state deadline for each objective to be reached by means of daily or weekly distributive schedule. In other words, goal indicates what the students will be able to do at the end of the course and objective, the method leading to its reach. It is necessary, extremely crucial to communicate the goal in the beginning of each course.

In addition to those, student should know the expectation in order for them to target it by means of a minimum of amount of study, which can be measured by their degree of intelligence and capacity of understanding based on how fast or slow they are when it comes to mastering a notion (Carroll, 1963; Hunter, 1994; Bloom, Hastings, and Madaus, 1971, and Engelmann presented by SRA/ McGraw-Hill, 2008). It would have been excellent for the teacher to tell his or her students the approximate amount of study time they should spend to master each notion. According to Carroll (1963), if they do not spend this allotted time to study, they are responsible for their failure as a problem of self-motivation may apply in that circumstance, which may need educational specialist interventions.

Why is it Important for teachers to teach this Way?

It help the students a lot in their concentration and focus when they know what their teachers expect them to know. Special details help the students know at what particular time to pay more attention to their teachers and to ask questions if they do not understand so they will make sure they do not leave the classroom without clear understanding. Later, we will see the danger of lack of understanding and this would be the reason why teachers need to ask questions on the content of the course which is a part of the curriculum.

3. Teaching Based on the Designated Curriculum

The teacher is obligated to teach the students based on a designated curriculum. Madeline Hunter thought that the first and greatest decision the teacher would ever take is what has to see with the teaching content. What Will I be teaching? All teaching models had to start with this question related to curricular content. This is what makes a big difference between Hunter and Bloom's models of teaching and learning. When Bloom focuses on textbook, Hunter puts emphasis on curriculum.

It's also clear in Hunter's head that teachers cannot choose their own and personal content; the teaching content choice is the direct and main concern of the system. Teachers need only to make sure they plan their courses based on the preexisting curriculum in order to teach what the students should know. If there is any textbooks, they should be written based on the curriculum and its concepts furnishing information which the students need to memorize and recall for necessary and crucial learning and which they also need to understand so learning can be effective for their greatest good. Therefore, improvement may be able to take place in the learning process.

Siegfried Engelmann put more emphasis on textbooks and capacity of the teachers to teach their course content as if they were empowered to create themselves that content or the curriculum. Authorities would need to be clear on the teacher's capacity to be able to make correction in the textbooks if there is any mistake in the textbooks. Nevertheless, they think that textbooks should not be imposed to them at 100 percent so they can correct mistakes they may find in the textbooks. Engelmann thought that, in this matter, teachers need to be boundary less. We may agree with Engelmann, but not with Bloom who put more emphasis on textbooks than the curricular content. The curriculum must be strictly taught based on available teaching strategies and methods in order to provide to the children necessary information, knowledge, and experiences to be stored for retrieval, in the future, for internal or external assessment.

4. Teaching to Help Students with Storing and Recalling

Storing information is so critical that learning cannot take place without it. Students need to learn things, to make some experiences, to have new information, and to store them in their memory system in order to retrieve them in convenient time need-based. If they cannot recall the information or a variety of knowledge (12) described by Benjamin Bloom in his book *Taxonomy of Educational Objectives*, then it is impossible for learning process to progress. According to Bloom (1956), information or knowledge should be recognized, stored, and recalled as the first step of the learning process cognition-based.

It implies that teachers would step intelligently and teach for memory. Before we state how they can do it, it is necessary to define memory and reflect on it to determine how it works. Memory refers to neuropsychological documents that human beings have capacity to retain in brain slate (Gardner, 1999); a region of our brain that can conceive time, past and future; we are aware of the past based on our neural state that creates what James (1950) called first effect of consciousness.

Human beings have capacity to keep information in their memory forever, which can come back at any time under no condition or circumstance. Sometimes, we associate ideas unconsciously and similar events may resuscitate stored information in the conscience. There is no memory without association or comparison of ideas. Retrieval is like a happening of cause-effect relation. However, it seems that we find ourselves, sometimes, in difficulty retrieving needed stored information. It is a brain defect, and abnormality, or a problem related to the process of storing information?

Putting information in the memory is nothing; retrieving it is the most important thing ever. We may not be aware of quantity of information that we put in the box; we will be aware of it and acknowledge that it has been stored whenever we retrieve it. Therefore, we have to have some kind of techniques to register needed information to be retrieved later. Retrieval will be easy or difficult dependent on the process utilized to put information in the memory (Gardner, 1999, James, 1950, Bruner, 1973, and Tileston, 1994 and 2004). James and Gardner even went far to say that we do not retrieve the same information, but its reflex by sensation effect. We do not want to develop this philosophy here.

Information retrieval has to see with stimuli and sensation. The nature and quantity of sensation and stimuli that motivate us to store information will make us able to achieve it in less or more time, and to conserve it for a short or a long period of time. This is why psychologists talk about short term memory and long-term memory.

It is more likely for a person to recall recent events with details than an event that took place a long time before. It is not enough to remember events' dates without details or to barely remember them to have just uncrystalized image of them. It would have been important to recall the shape and insights of past events whether they are recent or old ones which have happened long time before.

However, it happens that some children have difficulty recalling recent event, information that has been recently stored. Is it an abnormality or a problem of technicality, ignorance related efficient process of storing? Is it too in that perspective that Gardner (1999) classified memory as a kind of intelligence and organization? The more experiences we have of an event the more it is possible to retrieve that event with details regardless of time because we knew it as fact.

It is said that advised teachers are able to help students with memory problem. They just need to know what type of memory that is used in schooling system, which is semantic memory and which makes us consider teaching as an art. Semantic memory is defined by Howard Gardner as general memory. An intelligent use of semantic memory (place of words and documents) combined with lane memories such as episodic memory which allows to retrieve past information related to location, time, and content; automatic memory that helps retrieve condition response based on past experiences with the individual's effort, procedural memory which deals with muscle memory, and emotional memory to help with feelings and interests are all teachers need to leverage the capacity of the students to retrieve information (Tileston, 2004).

According to Hunter (1994), a teacher who uses regularly the blackboard, well position him/herself on the board, and who says things before writing them on board and erases previous writings before adding new important words to help the students with long-term memory. In addition, Madeline Hunter thought that rehearsal techniques utilized to store information can help as well.

To help students with short-term memory, according to Hunter, teachers need only to repeat the same information during a whole week. They may forget that information after they use it, she says. However, students may not lose the information for a very long time if the teacher repeat it for sixteen days. It is not necessary to repeat that information every day. It would be necessary to tell the information on the fourth day, eighth day, twelfth day, sixteenth, or more often.

Remember that memory is also selective; a person remembers information based on needs and importance. Therefore, it is extremely important that teachers give signals to the students about what they want them to remember, including words, declarative information. Mostly, teaching is done with declarative information to be stored in the sematic memory aided by lane memories that transform other kinds of information and the like (Tileston, 2004). At the last result, it is critical to help students with short-term and long term memories. A teacher who does not practice it does not possess the complete package of the art of teaching; they need urgent training.

Referring to our plan, it would be good to develop right now the fifth point of the testing strategy evoked above. But we should not go any far without reserving a little corner to test for memory and determine the kind of questions that have to be included in testing for memory as we have been developing our philosophy of memory and recall and what researchers have said about it. Now it is time to think of different types of testing based on Bloom's taxonomy of educational objectives.

Test for Memory

Test for memory should contain twelve types of questions if it's all right to follow Bloom's instruction on memory, basics of what makes learning possible. Let's see now twelve types of questions matching the twelve types of memory garnered from simple to complex questions.

1. Question Related to Remembering Specific Element

Based on Bloom's classification of educational objectives, memory refers to information stored to be recalled as the hard core of knowledge without reference. Questions asked to test this type of memory should be purely literal searching for specific things. If the teacher taught about a chapter in math textbook, he or she may ask question regarding a formula or what a particular element in a formula stands for. If a student remembers this critical element, it seems that he or she can remember the whole package because all the rest of the problem depends on it by interconnection. This is the first type of question that can be asked.

2. Question Related to Remembering References for Specific verbal symbol (Terminology)

A teacher may have taught what we call, in mathematics, ensemble. He or she might also disposed two ensembles with common elements in both representing their union that should be placed in a part of a diagram containing those common elements. Then that teacher could ask questions related to this part without the diagram or with it to be designed by the students if he or she has demonstrated it in class.

3. Question Related to Knowledge of Specific Fact

Bloom classifies specific facts as particular dates, events, persons, places, source of information, and specific facts related to culture, society (p. 201).

4. Questions Related to Ways of dealing with Specific

Here teachers can think about orienting the students on what they may have in mind when recalling specific information, according to Bloom's explanation. In this case, it is possible to ask a wake-up call questions that demand thinking and understanding data. We really don't know why Bloom inserts this kind of knowledge here. It would have been the last section of knowledge or memory turning to understanding or comprehension. Otherwise, treatment of specific retrieval had to be taught so any answer would be pure recall.

5. Questions Related to Knowledge of Convention

Convention is made by a group of people who are undertaking same activities to solve a problem happening in a special circumstance. To do so, it is possible that they create a special strategy which may include ways of doing things and steps and method facilitating that way. Teachers may ask students memory questions regarding this special circumstance if it has been taught.

6. Questions Related to Trends and Sequences

In this section, questions should be asked regarding the entire process of executing something from A to Z. That process may be recalled with the notion of time and context. If students remember the entire process, it is a sign that they have healthy memory.

7. Questions Related to Classifications and Category

This is where questions related to nature, similarity, and difference should be asked as long as those notions are taught. This question looks like, not a memory question, but intelligence one; anyone who can classify things and make difference between them is an intelligent man. However, they should be recalled before the classification process. Therefore, the art of classification must have been taught and learned.

8. Questions Related to Criteria

As criteria are conditional elements based on which knowledge, principles, or values are possible and acceptable for what they say they are, this type of recall is a little bit complicated. Questions related to criteria should be asked to check how healthy one's memory is.

9. Questions Related to Methodology

Teaching method is one of the highest subject matters that could be considered in any instructional curriculum. The systematic discourse made on method is called methodology which deserves brief explication here. It should not be extended to trouble the students' mind at a certain level of study. The level of method taught in classroom must be appropriate to grade levels students are in. An introductory course can be integrated in elementary and middle school where method could be defined progressively, and where kids could learn that method is the key of any achievement, including elementary tasks that should be realized at home and any things accomplished on the streets. Parents use a cooking method in the kitchen and cleaning method to clean not only the kitchen but also the entire house. They should be taught practically any particular procedures and processes. In assessments, they should be asked to remember method definition and any other procedure or process.

In high schools, teachers will make sure they teach the students about strategies of choosing appropriate methods to accomplishing particular tasks. Those are pre-colleges courses where the steps are introduced progressively. They include choice of topics, hypothesis, observation, collection of data, data analysis, verification of hypothesis, implication if applicable, and conclusion. Recall question must be also asked in this category.

Besides the notion of hypothesis, students must learn about assumption and about difference that should be established between assumption and the conclusive principle. Students should learn about all those and also learn how to make a conclusion being the end of the inquiry. Therefore, testing for methodology is a necessity (Bloom, 1971).

10. Questions on Knowledge of the Universals and Abstraction in a Field

It is difficult to determine abstraction in certain fields such as physics and biology. Anyway abstraction should be defined by thinking of something that is not present at the moment of the mental activity. The level of abstraction is characterized by the level of universality of a science, and universality of that science could mean that it may be practicable or applicable everywhere. In math, we can only see symbols and operate mentally by working with axioms and using the mathematical logic and reasoning. Teachers should teach kids how to do it. Have we ever met and seen a number, a triangle or a square? Mathematics ideas help us see and identify them in the reality. Therefore, mathematics is the highest abstractive sciences. Kids must be taught and assessed on these categories formally.

11. Questions on Principles and Generalization

The principles are what form the core of a theory. They can only be hypotheses or presumptions until proven tested and proven true knowledge in a given field. Generalization is another way to talk about universality of an object or a theory, and acceptance of a theory by the community of scientists. The students must be taught and tested on this category whether in high school or college, and at the university levels.

12. Questions on Knowledge of Theory and Structures

Bloom contends that the method used to put the principles together and the way of generalizing systematically the ideas form the structure of the theory. There is no science, no theory without systematization. This dimension should be explained and mnemonically tested.

Questions for Testing Memory on the Bell-Shaped Curve

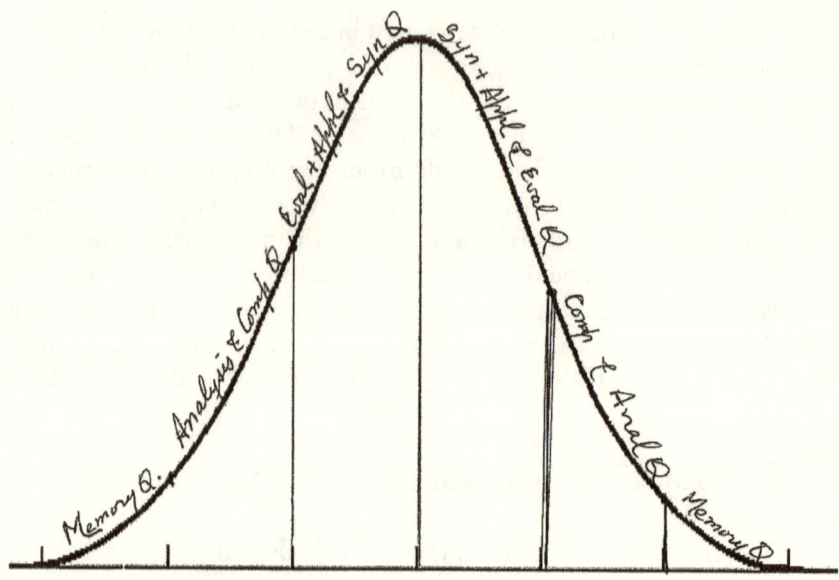

A total of fifty to fifty-two questions should be on this graph.
12

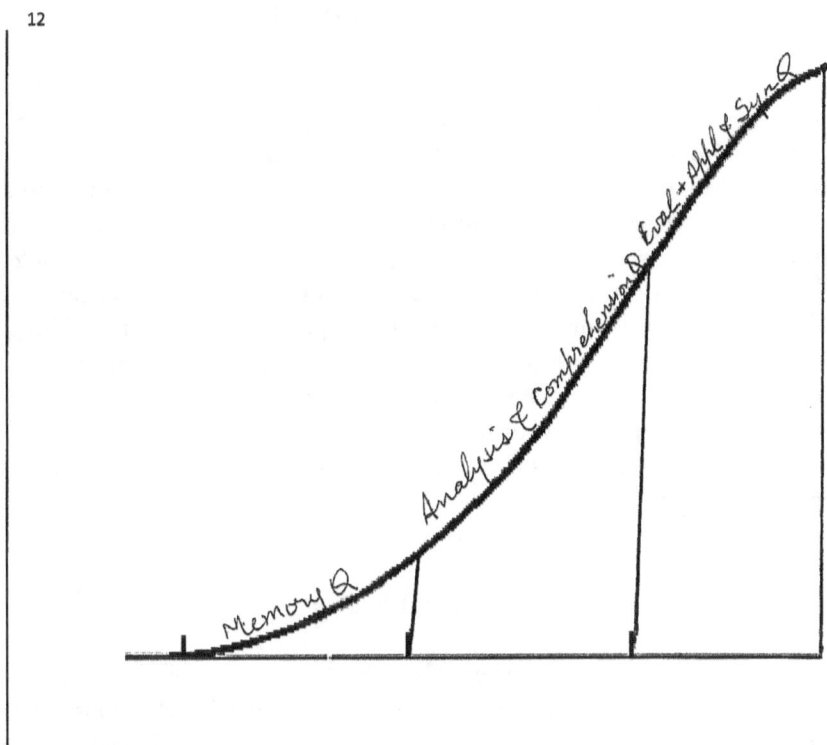

12

The first interval of the graph should contain memory questions (12).

The second interval contains analysis and comprehensive questions (3 + 3 = 6).

The third one will contain evaluation, application, and synthesis questions (2 + 3 + 3).

It implies that there are a total of 12 + 6 + 8 = 26 questions on the left side of the curve going from bottom to top, or from memory questions to skill ones. The right side that is not seen should also contain twenty-six questions from top to bottom or from skill questions to memory ones

Therefore, the schema should be replicate on the right side of the curve. However, one type of question may occupy all dimensions from left to right depending on the quantity of notions already taught and quantity of questions that should be asked.

Testing for Analysis

Remember that analysis was classified as the fourth educational objectives in Bloom's taxonomy because comprehension and application came before it. But we thought that people cannot comprehend anything before an analytical process whether it's done systematically or through a fast mental process of decomposing the object or the piece of transmitted information to see if that information is trustful and is understandable. This is the main reason why we place analysis as the second educational objective. Now it is time to test kids on this prospective educational goal. In our instructional strategy, we do not have testing for analysis and understanding yet. We need an analytical dimension of the encountering knowledge before we can ensure a degree of understanding.

The analytical process was completed when we analyzed the element, its relation with other pieces, and analyzed it based on a set of organizational principles. It is convenient to consider its following aspects:

1. Question on Analysis of an Element

An element is analyzed when it is broken down in its simplest different pieces. A necessary question in this matter would consider what constitutes physically that elements or what are its properties and or characteristics. Why this element comes to existence. The context in which it is learned and its importance in any given circumstance need to be considered. Can students do such analysis? One way to know it is via assessment. Questions should be asked consequently and appropriately on analysis of elements and so on.

2. Questions on Comparison of Elements

Comparison implies similarities or resemblance, or dissemblance, homonymous or synonymous aspects of two objects. An analogous method would have been able to help make a good comparison. Analogy and similarity that exist between two congruent ideas determine degree of similarity between them. If they are synonymous they may be employed or utilized in different contexts and circumstances. Be careful about this. If they are homonymous, they may have different meanings but same scripts and even same pronunciation. One may need only one side of the explanation to understand the nature of that element as he or she may need both sides because one idea may complete the other one to form a system of thought. This part of analysis should be taught and tested to know the students' thinking capacity or ability. They need skills to accomplish a complete analysis as they mentally grow up.

3. Questions on Analysis of Organizational Principles

It is about a systematic arrangement of ideas that helps make up connection between ideas and understand why this set of connection is made the way it is but not any other arrangements to make sense of what one wants to convey. It implies the logical aspect of the established discourse, which is nothing else than organizational principles. It is recommended to look for trustworthiness of this organization not only reality-based but also composition of ideas-congruency and harmony based. All these form what Bloom would call organizational principles that should be taught and tested to check students' skills. Knowing the secret of organizational principles is the key of making analysis possible. Upon the student knows the key words, what unites one idea to another one with their explicit contents and meanings, he or she can analyze any kind of organizational principles by decomposing them and giving them their exact value. This is now teachers can challenge their students.

Questions on Capacity of Analysis in Bell Shape Testing

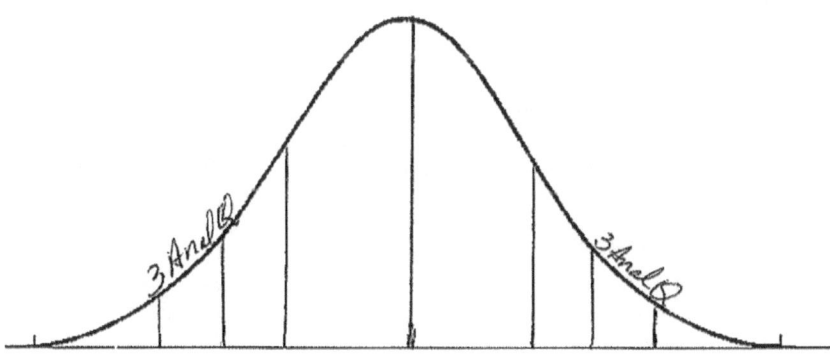

It is necessary to ask 3 analytical questions here. Questions move from less complicated to more complicated ones. It is going to be the same for all other types of questions based on Bloom's taxonomy of educational objectives modified a little bit.

5. Teaching for Comprehension or Understanding

We are back to our instructional plan and teaching for comprehension in Bloom's educational objectives. Remember again that Bloom (1971) placed comprehension secondly in his classification, but we place it thirdly for reasons that we explained before. Comprehension is the major element that teachers should put in mind because if the students do not understand the teaching content, they are wasting their time as failures. In that perspective, they will also fail the schools. Why do teachers teach? What is the main goal of education? The answer is to make students understand teaching contents in order to be able to pass internal and external tests.

The reading of Wiggins and McTighe (2011) makes us see a paradox when they contended that some teachers do not teach to make students understand their teaching, new concepts taught by them. Wiggins and McTighe thought that this phenomenon is astonishing because school makes no sense without kids' understanding. Worse is that those teachers to whom we referred before do not even care. Well, students' understanding is a necessary step in instruction. Teachers must care and show their interest in it in such away they should not teach new concepts until they check and test for it; it is the schooling process master key. Kid's understanding is the ticket to move forward.

Wiggins and McTighe (2011) contend that teachers will acknowledge the following things that students' ability in these areas:

- Capacity to explain
- Capacity to interpret
- Capacity to apply
- Capacity to shift perspective
- Capacity to empathize, and
- Capacity to self-assess (p. 4)

In Wiggins and McTighe's language, those capacities imply that students have to be able to perform autonomously meaning without their teachers' help, although it is not a simple recitation, but demonstration of their capacity of transferring what they have learned. The idea of transfer makes sense to us when we compare Wiggins and McTighe's approaches to that of Bloom. He thinks that students show comprehension when they are able to:

- Translate
- Interpret, and
- Extrapolate

The idea of translation implies that students would be able to say exactly what they see, hear, and read. However, some students don't know how to perform those steps in the comprehension process. They have to be taught primarily how to observe, listen so they can understand their teachers' expose, and how to read in order to be the most essential and practical teaching and learning processes. When students are able to perform those comprehensive things, it is more likely that they can go beyond information given. Whoever can do that can also show signs of understanding, sign of capacity to explain the content of their classroom lesson and to interpret whatever was taught to them. Therefore teachers must test for translation involving objective aspect of part-for-part of a communication (Bloom, 1971, p. 205).

However, translation is too literal and it does not allow students to further the sense of the information, and to discover what has not been seen, said, or read. The hidden aspect of them may be discovered at the interpretation level. Students would not follow the objective part-for-part anymore but they try to understand the piece of literature and speech being exposed or communicated. This is the gear of any interpretative activities?

What do I see, hear, and read? How can I understand what I just see, hear and read? Is it exactly what I just see, hear, and read? What is the meaning of this? Why is it like this but not like that? Is the context right? Does the writer or the speaker have something in mind that is not said? One needs to answer those questions in order to make sure that he or she does not miss anything that should be understood.

Because there is no other way for teachers to know if the students have capacity interpreting the information, they should assess them based upon. This is what theorists call "checking for understanding" (Bloom, 1971; Hunter, 1994 and 2004).

Lastly, Bloom thought that students must be able to extrapolate. Extrapolation is effective when the students are capable of making inference, implications, and determines consequences of the information communicated and read. Teacher must assess students for this level needing special skills and abilities. We refer our readers to the graphs above to see where those three questions are placed on the bell-shaped curve for the progressive structure of the test.

6. Using Feedback Regularly to Check for Understanding

Checking for understanding is what Douglas and Nancy (2007) called confronting misconception such as incorrect belief. Students, sometimes, think that they understand when they really don't. It is critical to sincerely check for understanding and make sure they do understand in order for them to progress. No progress can be made without students' understanding. Checking for understanding, said Douglas Fisher and Nancy Frey, is a systematic approach to formative assessment (p. 3). It is important that feedback comes immediately after formative assessment.

Feedback is what is performed after or while a teacher is checking for understanding in order to tell the students how they are doing: good, bad, or between good and bad. An excellent student must do only the good, but neither the bad or nor even average grade. An average student earns 70 percent of the total possible points while good students earn only 80 to 85 percent of possible points (Carroll, 1963; Bloom, Hastings, and Madaus, 1971; and Engelmann, 1991). If a student earns less than 80 percent he or she should be given some special feedback and be receiving corrective instruction. This is where teaching is individual, but not group activities anymore or can be intertwining with group study.

Feedback is presented to students in two ways: verbal and notes or written remark in the margin of the paper. Feedback should be clear and appropriate, and specific (Brookhart, 2008) so teachers would say exactly what students need to know, the errors, and the missing points. The nature of the test for which feedback is provided is what Scriven, cited by Linda Allal and Mottier Lopez of Geneva University in 1988, and Bloom (1971), called formative assessment. Brookhart has provided four indices to recognize good feedback:

1. Written feedback for comments that students need to be able to save and look over.
2. Oral feedback for students who don't read well.
3. Oral feedback if there is more information to convey than students would want to read.
4. Demonstrate how to do (deal with feedback) if the student needs to see how to do something or what something looks like (p. 17).

Now the best a teacher could do is ensure that students understand his or her feedback and are ready to improve and make a difference. This must be done after each session or very often. Two best ways to do it: ask the students if they have question; if they do not have one, then ask them some important ones, or do the next formative assessment. It means that must teachers test for feedback, and this part must be internal only; understanding feedback may not be a part of the curriculum; it is the teacher's concern.

7. Assessing the Students Based on the Teaching Content

Robert Marzano contended that Bloom has highly influenced the people, educational theorists and leaders, educators including principals, teachers, and university professors when it comes to considering his *Taxonomy of Educational Objectives*, but he has been scold and scorned for not putting emphasis on the curriculum. Content is the centerpiece of all instructions setting.

We have talked about teaching content or curriculum a lot before; this is an extension of thinking curricula. But it is necessary to say it here that curriculum is where educators set to meet educational objectives, where subject matters are found, and also where teachers and assessors find their test content. If any exogenous questions are asked, then it should be called mental torture, educational and intellectual abuse destroying the students' intellect, and it should be called silent and slow educational killing process.

8. Using Cooperative learning

It is widely recognized that cooperative learning is an instructional theory that was developed by David W. Johnson, Roger R. Johnson in collaboration of a contributor whose name is Edythe J. Holubec. However, they usually refer to Johnson when it comes to paternity of cooperative learning. It is an honor to state here that Bloom started his students' learning process with group study too, but not as systematic as Johnson. Vygotsky put some interest in group leaning to help his kids benefit what he called zone of proximal development.

According to Mariane Hedegaard who published an article, in 1990, on Vygotsky's proximal development titled "The Zone of Proximal Development as Basis for Instruction", he presumed that group of study is extremely critical not only for the development of an individual in general, but also for students who want to learn and to increase their capacity of understanding new concepts that appear to be more difficult to be grabbed alone. This idea is considered as basis of Vygotsky's instructional theory. Here is what Hedegaard reported regarding Vygotsky, which we want to transmit textually:

> The child is able to copy a series of actions which surpass his or her own capacity, But only within limits. By means of copying, the child is able to perform better when together with and guided by an adult than when left alone. (p. 349 of Vygotsky And Education: Introduction and

Applications of Social Historical Psychology, Edited by
Louis C. Molly)

Molly (1990) developed a similar and congruent approach when he
presented Vygotsky's zone of proximal development as a social system.
Molly called upon Vygotsky to say that "maturing or developing mental
functions must be fostered and assessed through collaborative, not
independent or isolated activities" (p. 3). He further cited Cadzen (1981)
who considered performance before competence where "the zone makes
performance possible" (p. 3). Later, in the same text (p. 9), Molly came with
the same idea where he used the exact word (cooperation) that Johnson and
Johnson would use in 1990.

Vygotsky was not a speculator, but a researcher; he discovered the
foundation of his concept of zone. He conducted a research on learning
in a Danish Elementary school. He was supported by a group of teachers
of biology, history, and geography. The participants included from third
graders to fifth graders.

Johnson, Johnson, and Holubec (1994) thought that it is necessary to
put fast pace learners with low pace ones to facilitate equal distribution of
learning. It is more likely for low pace learners to grab difficult notions
from fast pace learners than from the teacher. Therefore, the desired
cooperative learning group would be heterogeneous, including students
with different background abilities to expect greater outcomes.

What about the notion of gifted students today who benefit special
classes offered by the system? How would the system hold back low space
students to access the best that they offer fostering and promoting high
standard education to all? Justice for all cannot stand properly without
education for all. Anyway, they can be forgiven because of their limit
regarding education and environment and plasticity of the brain.

Johnson contended that learning is something that students do, but
not the teacher. Teachers need not only to set the groups with desired
size (two to four), but also to create rules to make sure cooperation really
exists among the students, and that the teacher supervises them in case
they need help.

What are the criteria based on which one can identify a cooperative
learning group? The group should be assigned with assignments that
target the educational objectives after choosing the instructional
material. In addition, roles should be given to each member to facilitate
individual accountability; the instructor would make sure there is positive
interdependence between them so they can share their experiences using

face-to-face and promoted interaction to benefit the entire group. This can work when members encourage each other's efforts.

How to evaluate group works? Even though there will be individual contribution to help attain the educational objective, the group should come out with only one homework to be evaluated (Johnson, 1994). However, Bloom thought that students should be assessed individually so the teacher can instruct particular students with needs of extra explanation. This kind of instruction would be one-to-one. We think that a teacher may use both method of evaluation and provide a grade for each activity. There will be two grades that the teacher should add together and divide the total by two in order to find the final grade for each student.

Which assignment should have a greater percentage? The teacher will decide this him/herself. Flexibility can allow him or her to integrate the students in that decision making process for their greater involvement and satisfaction.

9. Assessing Regularly the Students Using Formative Assessment Method

We do not have to say a lot here about formative assessment because it has been developed enough in the first part of the research referring to different theorists such as Bloom (1971), Black (1999), Guskey (2007), and Marzano (2003 et 2007), etc. All those practitioners recommend the practice of formative assessment as often as it is needed, especially in the beginning to find out backgrounds of each students, after each class, and before the summative assessment.

When it comes to formative assessment methods, it seems to be clear that teachers need to use the bell-shaped curve testing structure going from memory questions which are simple questions to skill and complex ones. It is not recommended to ask complex questions first and simple ones next; this system of assessing is discouraging and lessen kids' disposition to take and pass the text. Student need to feel a little bit confident in the beginning of the test and catch some strengths to be able to think when complex and tough questions appear.

This method is used progressively because, there will not be complex and difficult instructions in the beginning, and a teacher cannot assess the students on notions that are not taught yet. Teachers should not be tricky and should not want to catch students being wrong by guessing on any answer; teachers instead should be happy when students are highly confidents on test questions and are highly motivated to increasing up their performance, and to score higher on high-stake-tests.

It would be all right to assess the students on the final summative assessment by asking questions that look similar. Be careful not to ask the exact or same questions in that final because, sometimes, students steal that exam and work on it during the entire semester and get advantages on those who did not steal and who are innocent. Therefore, ask similar questions, but not identical ones.

After this trial, teachers should instruct the students to attain all educational objective that Bloom has proposed in his Taxonomy of Educational Objectives and assess the students on them (from information or memory, or knowledge to judgment) based on the course content or the curriculum (from simple notion and skills needed to higher-order notions and skills needed).

Lastly, bell-shaped structure will be completely used in one course depending on the number of tests a teacher plans to give to his or her students. At least, it would be necessary to perform two bell-shaped tests as formative assessments in a single course before summative assessment takes place to allow students get more trials and have more teacher's corrective interventions.

The students could have a better ideas of that testing system if the teacher followed the bell-shaped structure after completing each objective. To holistically get this package one should refer to Bloom's first educational objective slowly and try to understand and practice the second one which would be comprehension or precisely analysis based on change that has been made by us. Teachers should dispose adequate quality time to make it happen. It is going to pay off.

Following objectively the BST is critically and extremely recommended. We are pretty sure if any teacher steps up and makes this method of testing students his serious business affaire his or her students will succeed at 80 to 99 percent. All students can and are able to learn without exception if they are taught excellently to repeat Carroll, Bloom, and Engelmann. Therefore, students' failure belongs to their teachers if there is no mental inconvenience or retardation. If there is any, then the staff administration should know about placement programs and how to facilitate instruction for this category of students and make things easier to benefit the whole educational system.

Putting apart this special group of students who need special education, school administrators should develop entrance tests capable of indicating in what class each student should be objectively placed by determining what they can actually do alone and what they can do with a teacher's help or peers' help so they can be instructed based on Vygotsky's zone of proximal development. This placement test should not be organized based on bell shaped testing because they are not taught yet. Bell-shaped testing should take place when teaching and learning process begin.

CHAPTER VII

Organizing the Research on Bell-Shaped Testing

Organizing a research has never been an easy task. Researchers can be struggling undertaking it very seriously because there are a lot of things that are required to make a research paper readable to the readers and acceptable to the community of researchers. To be recognized having these qualities, the writer should be able to prove that there is great needs of any nature in a domain of knowledge whether social, economic, scientific, educational, etc. and his or her research comes to fill up the empty space terrifying the nature.

It is exactly the case of testing system in educational matter where parents are receiving unsatisfied reports for their kids who themselves do not know anymore what to do to have satisfactory grades for courses taken everywhere including elementary schools, middle schools, high schools, colleges, universities, and technical schools. These educational thresholds have been traumatizing for so long our societies (USA) and the world in general. It was time for someone to come up with a solution after diagnosing and analyzing the situation with sense and expertise-like.

It is not without knowing that the writer should use logic, clear and academic language sometimes with a mélange of simple and rhetoric styles to be elegant and to tap the readers' reading appetite. In addition, the content of the paper should not be light and that the writer should have the gut of matching the style utilized in order to be able to lead the reader from the beginning to the end with open mouth, astonished to never want to abandon the paper when it comes to placing it, one could have said, at night, on the bed head. Nevertheless, the writer would have to create or select a good research method under an infallible methodology.

Methodology

Best of all, the writer would have to create a research method or to use an existing one that is working providing means of assuming what appears to be true. That assumption would be considered as hypothesis revealing the writer or the investigator's worldview, one's perception and conception of life, which may be verified and generalized after proof of evidence that what has been thought was so (Rene Descartes). Creswell (2009) defined worldview as "a basic set of beliefs that guide action" (p. 6), and based on these beliefs, the researcher will choose a method which could be quantitative or qualitative research methods, or a mixed research method.

Based on the nature of this particular research, it is necessary that we use a qualitative research method (QRM) which will allow us to express freely our view and explain it according to our observation, comprehension of what we see and feel. Qualitative research method will also allow us to seek other people's opinions by interviewing them or having them answer our survey questions. This is how, we think, that we may be able to gather data and analyze them, and search for their validation in order to arrive at a useful and transferable conclusion. Analysis of the data may also lead us to the use of Glaser's discovery of grounded theory method which is an extension of the qualitative research method. The purpose of it is to organize the data by using symbols as codes, tools for a better classification of types of data needed, using different approaches, and, at last, increasing our knowledge in comparing groups of data and cases (Glaser, Anselm and Strauss, 1967; Heath and Cowley, 2004).

Hypotheses

H1: If the teachers teach based on the established curriculum and use a good instructional method, students who are tested only based on what they have learned from their teachers and also based on the bell shaped testing approach—where questions are arranged from the bottom to the top, from simple to complex questions, or from retrieval to synthesis questions—will always earn grades no lower than 80.

H0: If the teachers teach based on the established curriculum and use a good instructional method, students who are tested only based on what they have learned from their teachers and also based on bell-shaped testing—where questions are arranged from the bottom to the top, from simple to complex questions, or from retrieval to synthesis questions—will not always earn grades 80 or above.

Assumptions

Based on the instructional reviewed literature and our given hypothesis, and our strong instructional beliefs, we assume the followings:

1. It is extremely critical that teachers teach the students based on the established curriculum and that they follow it strictly so state curriculum content be aligned in order to make the students very successful.
2. It is extremely crucial that teachers teach based on an instructional strategy and a teaching method to make good curricular unit explicit and understandable. Otherwise, teachers are not only wasting their time but also the students'.
3. The students will be successful and pass the tests above average if the previous principles are highly applied and if the test questions extracted from the programmed curriculum.
4. The students will be able to pass the tests with 90 and above close to 100 percent if teachers strictly follow the bell-shaped testing system procedures.

Corollaries

1. Students need adequate time to be taught and to study out of which no success is possible.
2. Individual remediation and corrective actions are necessary in such a way make up can be possible to lead the students to success.

Recalling the Problems That Were Identified

Problems have been identified as threats of the educational system hindering the students' success because of lack of curriculum alignment with the materials used by the teachers and their insufficient use of method of teaching to make kids understand very well and get insight of the lessons. The system is stalled and struck. On the top of that, teachers do not align their teaching to the assessment and test questions. Lastly, teachers and state assessors do not structure the test questions and make them comprehensive so test takers could psychologically gain some assurance that they have a chance to pass any given test within the teaching context. Those allegations need to be verified through empirical research so formal and final recommendations can be made.

Getting Ready for the Empirical Research

Empirical research (EP) refers to deep reality that we are living as human beings, observations that we make as researchers and investigators, data that we collect from our past and present situation that we seek to understand, and experiences that we also make to verify what we have said a priori as assumptions in order to construct a scientific theory or to support another one in its eventual propensity toward generalization for the good of the humanity. In our case, we are just eager to find a solid foundation for the bell-shaped testing theory.

We are striving to be a little bit empiricist, but not fundamentally positivist as empiricism is one of the positivism's fundamental aspects. We agree that a scientific thought tends to be more objective than suggestive, but not in the senses of B. F. skinner, founder of the psychologism and the behaviorism and in the sense of August Comte, founder of positivism based on which "knowledge of the world can be acquired through the senses and experiences" only (Lunenburg and Ornstein, 2004, p. 39).

Our point is not different from that of William James in his pragmatic philosophy. He claimed to be empiricist because pragmatism finds its roots in the reality. However, he also claimed that his empiricism has nothing to see with radical empiricism developed by John Locke, George Berkeley, and David Hume. Here is what James (1991) wrote: "The latter (radical empiricism) stands on its own feet. One may entirely reject it and still be a pragmatic" (p. 4).

We are empiricist at a certain point just to be scientific, but not a radical empiricist. Radical empiricists such as those mentioned above, including Comte and Skinner are sick and they slipped away from any curative diagnosis. They missed the reality being astray and strayed away from them unconsciously. They lie to themselves as they cannot see and touch what is inside of them that made them think the way they thought. Plato, the father of idealism had said that this is idea that leads the world but not the material things. Charles S. Peirce, introduced by Justus Buchler in *Philosophical Writings of Peirce*, published in 1955 while the original was published in 1940, is recognized to be the founder of the new and contemporaneous philosophy called pragmatism, which went further to introduce a kind of idealistic experimentation based on deductive reasoning.

He placed his a priori method over scientific method by declaring that "the a priori method is distinguished by its comfortable conclusion" (p. 20). Buchler (1955) was enchanted and expressed his enchantment for Peirce realism. We appreciate Charles Peirce for his firm conviction in his presentation of his pragmatic method laid on continuity method of discovery the truth as we also appreciate the founders of the rationalistic foundationalism of the modernism (Rene Descartes and his disciples who are the radical empiricists.

To go back to our intention to get into empirical research, we agree with Cone and Foster (2006) who assumed that we could be also away from constructing a scientific research if we do not test any relationship because "science is the study of relationships between variables" (p. 68). We are actually trying to establish fundamental relationships that should exist between present curriculum (independent variable) and test questions (dependent variable), between teaching strategy and method (independent variable) and students understanding (dependent variable), and between alignment of curriculum with teaching content (independent variables) and test structure also (independent variable) with students' ability to pass the tests with higher grades (dependent variable).

It is possible in this case to use a kind of discovery of grounded theory which requires, based on Lather's approach (1986) cited by Creswell (2009), to search for reciprocal relationship between data and theory. We are stepping ahead to build up our questionnaire in order to gather pertinent data so we can establish that relationship if it really exists and which can help see and understand clearer our theory after an exhaustive analysis. As a matter of fact, we plan to meet with fifty students from different grade levels and fifteen to twenty teachers teaching different grade levels in Jacksonville, and in South Florida. We predict, a priori, that there are rich relationships between curriculum and teaching content, between outstanding teaching and students' understanding, and between test questions and structure of the tests and students ability to pass those tests with higher grades.

From the data that will be collected, we will try to build that theory mentioned above, the discovery of grounded theory which cannot be falsified neither demarcated (Popper, 2002) because it is going to emerge from the data (Glaser and Strauss, 1967). We could read it and verify it ourselves from Glaser and Strauss (1967)'s book where they wrote what follows: "Theory based on data can usually not be completely refuted by more data or replaced by another theory. Since it is intimately linked to data it is destined to last despite its inevitable modification and reformulation" (p. 4). We will enjoy adding this wonderful theory to those we have already used in the gallery of the in-use-theory when it comes to instruction and our own instructional theory (Argyris, 1994). The most important thing here is that we know why we use them and when to use them. Now, it is time to seek the data we've been talking about for so long.

Questionnaires

We say questionnaires because there will be two sets of questions to test at different levels: one for the teachers and one for the students. It means that some data will be collected from each of them in order to have a better view of what teaching and learning are all about and what kind of instruction the students are receiving from their teachers. Is it adequate or not? Are the kids satisfied or not? Anyway, different points of views may be helpful in this case. The students' questionnaire will come first to be followed by the teachers' one.

A. Questionnaire for the Students

1. Do you think that topics developed by your teachers in the classrooms are identical to topics mentioned in the curriculum imposed by the school board?

2. Have your teachers ever presented to you goals and objectives of their courses and allowed you to discuss them given you some time to discuss anything related to them?

3. Do regular tests you usually take match what you are taught by your teachers in classrooms?

4. Do your teachers provide feedback to you regularly and undertake corrective actions or reteach the course that you did not understand?

5. What are the first questions you usually encounter in your tests?
   a. Easy questions
   b. Memory questions
   c. Tough questions that demand application of skills?

6. How do you feel happy or more comfortable to take the exam when tests begin with easy or recall questions?
   a. You don't like those tests.

    b.   You feel psychologically more ready to concentrate and pass.

    c.   You feel discouraged and you lack concentration.

7.   How do you feel when tests begin with tough questions?
   a.   You don't like those tests.
   b.   You feel psychologically more ready to concentrate and pass.
   c.   You feel discouraged and you lack concentration.

B. Questionnaire for the Teachers

1.   How often your course content meets your state established curriculum requirements?
   a.   Always
   b.   Sometimes
   c.   Often
   d.   never

2.   Have you ever received a curriculum from your school administrator?
   a.   Yes
   b.   No
   c.   Sometimes

1.   Do you use an instructional method in your classroom?
   a.   If yes, then what is it?
   b.   If no, why?

2.   How often you check for your students' understanding?
   a.   Always
   b.   Sometimes
   c.   Often
   d.   Never

3.   How often you perform formative assessment and provide feedback to your students?

4.   What do you think about formative assessment?

5.   Do you use to start your assessments with easy or memory questions or tough ones?

6. How are your student scores when you start assessments with easy or memory questions?
   a. Higher
   b. Lower
   c. Average

7. How are your students' scores when begin assessments with tough and skillful questions?
   a. Higher
   b. Lower
   c. Average

8. How often you provide feedback to your students and what does it mean to you?

9. How are your students' scores when begin assessments with tough and skillful questions? a. Higher
   a. Lower
   b. Average

CHAPTER VIII

Collection and Analysis of the Data

It is good and wonderful to arrive at this critical point where obligation is imposed to us to gather data in the objective to prove our hypotheses and respond to pertinent questions surrounding the hypotheses and conjectures. Data are what pretend to give these presumptions and hypotheses a scientific aspect. We are not here to create data, but to gather them based on special occasions that give us opportunities to review personal experiences that we have made as former students, teacher, and previous observations which may be of need in this research. Based on these strategies of collecting data, one does not need signature of any participants for consent form.

Another type of data collection strategy (DCS) will not require consent form. It is collection of data from survey where participants are freely giving the data by responding to some questions pertaining to a research topic out of any special settings. It means that we did not need to see the participants face-to-face. Those participants are professionals such as nurses, and teachers who shared with us their past and current experiences. We also met high school and fifth-grade students who responded to our survey questions.

It is just good to inform that we haven't asked to provide any consent form which was available. The fact is that we had not to perform any experiments and perform anything that could put lives in danger psychologically, mentally, and physically. Now is it about time to consider individual source of data and the data collected from them?

The Researcher's Past Experiences as Primary Data

As a former student, the researcher has never been told about the main goal and objectives of his courses. Some teachers have tried, but they have not put emphasis on critical notions that needed to be mastered and how to master them. Unfortunately he grew in a tradition where teachers mostly promoted recall, but a little comprehension. It is not that they did not teach effectively, but there have been a lack of teaching method and learning strategies to enhance learning as hunter expressed it in her book, Enhancing Teaching, published in 1994 promoting mastery teaching which almost meets Bloom's mastery learning requirements.

In addition, the investigator has not been taught about types of assessment which are formative and summative assessments (F&SA) and how to get ready for each of them when approaching the judging moments. He reported that those were dark moments and he was left out to guessing. However, his test questions tended to be memory questions putting apart the problems which demanded deep reflections in mathematics, physics, chemistry, and so on. Fortunately, teachers gave him a lot of examples, and the students who wanted to be rewarded with excellent grades gave themselves sufficient time to form group studies to work out not only those problems but also many other similar ones.

The researcher reported that a few teachers provided feedback to the students, but not as individual the way individual learning intends to be under Carroll (1963), Bloom (1971), and Engelmann (1986)'s learning theory. Consequently, he affirmed that he couldn't lie about him not being able to be an A student. He used to be a B student; that was not too bad, he said although some C grades used to come around troubling his glorious educational success path. Lastly, the researcher reports facts related to observations he has made while he was taking his instructional leadership test months ago in Jacksonville, Florida. It was not a good experience at all because that test was so long containing 195 tough questions only. It required too much concentration. The researcher observed by this time a girl next to him who was taking the exam. She could not face it and left after a short period of time, and it seemed that some other test takers followed her. Some others stretched a lot before they could take it to the end as did the researcher himself.

It turned to be a failure.

Data Collected from the Participants

We have to admit that this research was not easy at all, especially when it comes to data collection (DC). The greatest challenge that we encountered was to meet the people and get them fill out the surveys so we could have their entries and beliefs regarding the Florida testing system which needs to see with quality teaching and learning. We realized that everybody was eager to know what holds the system down so long and causes the students fail to earn satisfactory grades, and this is the reason why we could collect some valuable data with small sampling sizes for two categories of participants (COP): students and teachers.

The population, as said before, was Florida represented by four cities, including Jacksonville, Orange Park, Fort Lauderdale, and Miami. We mostly met high school and fifth-grade students, elementary teachers, and some professionals. Why didn't we pick students in lower grades? We thought technically that they would encounter difficulties understanding the questions. Anyway, that was a great experience.

Thirty students and twenty teachers participated in that educational research. It means that sampling size for students was 30 and sampling size for teacher was 20. We did not help them answer the questions. We let them do it themselves based on their experiences with learning, teaching, and testing. Most of them answered yes or no and added valuable, incredible comments that one would never imagine for open-ended questions. The most important is that they made comments for simple questions designed to receive a simple yes or no.

Now it is important to debrief the students' questionnaire and consider the answers with some statistical aspects. Let's, first, group thematically questions that are similar together using different letters as symbols in order to foster a fast and an accurate analysis of the data:

A. Similar questions on the working curriculum (WC) or teaching content (TC) and direction provided to the students (questions #1–3).
B. Similar questions on the beginning of the tests (BOT) (questions #5–7).
C. Question regarding teacher's feedback (TF) (question #4).

Notice: Questionnaire for each group of participants (students and teachers) is divided in different categories based on themes developed in it. It means that we put themes that are similar together.

A. Similar Questions on the Working Curriculum or Teaching Content and Direction Provided to the Students (Questions no. 1–3)

For questions related to curriculum in the first group which is called again (students participant A, or SPA), it is found that 88 percent of students are satisfied with instructions received in class. It is also found that their teachers have presented their working plan (WP) and allowed them to discuss it based on their education needs. Lastly, it is reported that their teachers have always assessed them based on the curriculum and class discussions (C&CD).

B. Similar Questions in the Beginning of the Tests (Questions no. 5–7)

However, only 67 percent of student participants in the second category or B category (SPB) only felt that their assessments often begin with memory or easy questions which usually help them to courageously take the exams with a winning mind from beginning to the end. On the contrary 33 percent of students felt discouragement and disenchantment. It means that their tests always begin with tough questions, and they waste all their times concentrating on them to finally take a chance with guessing which has never been a good sign for them when it comes to thinking about passing grade.

C. Question regarding teacher's feedback (Question no. 4)

A total of 55 percent of student participants in the third category or category C (SPC) declared that they received feedback from their teachers; however, 29 percent of students who answered to the survey questions declared that they never received feedback from their teachers. Unfortunately, it seems to be strange that 5 percent do not event know the meaning of feedback in respect to classroom instruction (CI) because they have provided no answer for question number 4. It may be clear that their teachers have never mentioned this word because they had no intention to provide feedback to their students which deems to be critical for a profound educational research analysis.

As this part marks the end of data collection related to student learning experiences, it is time to debrief teachers' questionnaire and answers where more exciting data may be found for the richness of that inquiry.

Debriefing Teachers' Questionnaire and Answers

There are four groups of questions found in the teachers' questionnaire that are arranged in the following ways based on their similarities which should be debriefed separately. Those are:

A. Similar questions on the WC (questions no. 1 and 2)
B. Question related to method utilized (MU) in the classroom (question no. 3)
C. Similar questions on checking for understanding (CU), formative assessment (FA), and feedback (questions no. 4–6 and no. 10)
D. Similar questions on the BOT (questions no. 7–9)

A. Similar Questions on the Working Curriculum (Questions no. 1–2)

A total of 91 percent of teachers declared that they always receive a curriculum from their school administrators and they have never missed an occasion to match it with their TC, which they always present and discuss with their students. What about instructional method (IM) that those teachers use in their day-to-day practice?

B. Question Related to Method Utilized in the Classroom (Question no. 3)

A total of 91 percent of teachers used a teaching method to transmit knowledge to their students, and they conveyed that it is of great importance for any regular and normal teacher who wants to be successful in this field. Most of the teachers revealed their teaching methods which are the following: whole and small group intention (WSGI), visual aid (VA), gradual release (GR), and computer center independent (CCI) (individual).

C. Similar Questions on Checking for Understanding, Formative Assessment, and Feedback (Questions no. 4–6 and no. 10)

Ninety-one percent of teachers affirmed that they check for students' understanding (SU) regularly, practice FA very often, or after each session, and provide feedback each time it is necessary. According to those teachers, there is no valuable teaching without these elements which constitute the heart of instruction. Unfortunately some other teachers (9 percent) missed this noble opportunity to instruct their student efficiently by using these educational and technical tools.

D. Similar Questions Related to the Beginning of the Tests (Questions no. 7–9)

Teachers are split when it comes to type of questions they should put in the beginning of the tests which should be whether easy or memory questions, or tough or difficult skillful questions. Some of them (10 percent) left those questions without answer meaning that the kind of question with which the test should begin does not matter. However, some of the participants said that although they start with easy or memory questions, their student still don't make higher grades, but average. Lastly, 10 percent of participants informed us that all their test questions are just tough and this is how exams should be conceived. They seem to be wrong.

Data Analysis

In general, analyzing something consists of an intellectual effort of knowing this object inside out. It is a way of dissecting it in its different parts to understand its specific aspects. However, profound analysis requires the analyst to regroup the analyzed elements in patterns based on their meanings. It will be necessary to check out the language and rhetoric that the communicator has utilized to convey what he or she wanted to express. It is in this perspective that Hunter saw analysis in the following ways: "analysis may be accomplished using language, sonorous cues, or visual space. To analyze you must be able to think categorically to organize and reorganize information into categories" (Hunter, 2004, p. 106). This approach is not different from Glaser and Strauss (1967)'s view of organizing information in patterns, basis of our data analysis activities with intent of discovering new knowledge. This is what we did in debriefing the data that we have collected.

To Bloom, Hastings, and Madaus (1971), analysis deals with breakdown of communicated element in its constituent in order to clarify the communication: the way it is organized and managed to convey its effect. Those are what we intend to do in this research in analyzing the participants' answers composing the data that we have gathered. It is worth analyzing the two groups of data (those of the students and of the teachers) separately as the former is regarding learning and the latter teaching. After that, it is imperative that we search for a common interest binding teachers and students in an intellectual package, and it may be a way of creating two minds for the future in the sense of Gardner (2006) revealing five minds that the future will need of which creating and synthesizing.

Analyzing the Data Regarding the Students

Eighty-eight percent of SPA have expressed their satisfaction related to how their teachers use the state curriculum to provide them with valuable information or knowledge they need to be scholarly successful. It is that the teachers allow them to discuss democratically the content of the courses. Two educational strategies would be interestingly marking the competencies of the teachers. Those are the use of curriculum and dialogue with students to make sure that their instructional needs are met. In addition to those, teachers assess them based on the curriculum and discussion occurred in class. It is a great sign that those teachers teach based on standards and make it easy for the students to be successful.

It is extremely critical that all the students be successful because success of a student indicates that teachers are highly qualified, the school is doing a great job, and the educative system is up to date and excellent. It simply means that the success of a single student, a kid is the success of the entire system. The opposite is also true. Therefore, teachers are in obligation to teach at the level of excellence. Why would a teacher like to fail any of his or her students? I remember that, at Florida International University where I studied Social Work, my last class professor tried to fail me because I reported her for wrongdoing, racial discrimination; she was not fair at all practicing that kind of discrimination. A teacher should be ethical minded, impartial, and should see all the students as human beings equally likely.

It is said that three basic strategic teaching elements are necessary to lead the students to success: teaching based on the established curriculum, presentation of the WC to the students with the course objective and allowing the students to discuss the plan, and testing them based on the curriculum and the occurring CD. As 88 percent of students expressed their satisfaction when it comes to those three points, it is positive for the system; however, they are not enough. In order to assure the expected educational success, they need support of some other fundamental teaching strategies (FTS) such as creation of what we call structure of the tests, and the types of questions asked in the BOT which has a lot to see with the students' mind and emotions, center point of the next group of questions.

Sixty-seven percent of students said that their tests always begin with easy or memory questions. Memory is defined as capacity of recalling with fidelity and exactness the information displayed before. It is necessary that students possess this capacity because there will be no exercise or demonstration of skills without this first step of acquisition of knowledge, there can be no scientific experimentation, no analysis, no comprehension, no explanation, no synthesis, no extrapolation, and no judgment. In addition, psychologists agree to say that human being is out of sense without this important brain device which is memory. Therefore, teachers are wrong when they think that it is not worth asking memory questions. It is unfortunate that only 67 percent of teachers understand this dimension of education and are not ready to understand it deeper.

The difference (33 percent) forms the category of teachers who need to understand that memory is in the heart of teaching because it is at the basis of the cognitive process. Therefore, it should be checked out in order to ensure that students have what they need mentally in order to learn properly. Three reasons for which teachers should test for memory: check if students have paid attention to what have been said in the classroom, if they have studied (rehearsal activity), and if they are experiencing any mental problems, or memory deficiency. Any teacher who misses the opportunity to test for memory based on those three checkpoints also misses the opportunity to be a successful teacher and to lead the students to success.

Another reason for which memory questions should be given in the beginning has to see with psychological effects. Those students who agree that memory questions make them feel like they are ready to courageously take the test with a winning idea in mind, expressed true heart feelings. They might be definitely right. It means that when the first questions are tough they are definitely discouraged spending all their energies in questions they will never be able to answer. It is more likely for them to fail. Teachers need to have in mind that failure of a kid or a student is the failure of their own, everybody and that of the system.

We do not think neither that the students have any satisfaction regarding feedback. Fifty-five percent only informed us positively about feedback. Imagine that 45 percent of the entire class do not receive feedback from their teacher. What would it be if it's not a catastrophe? Feedback is one of the greatest learning tools when checking for students' understanding. Students' understanding would be the teachers' major goal in their teaching activity. It's not only those who earn higher grades who deserve feedback, but all the students and especially those who could not make it happen. Attention must be focused on them. On the contrary, the results will be catastrophic.

Synthesis for these categories could make observers and stakeholders understand that it seems that the system fails to provide standards education to the students in order to improve and to be successful. It is to be remarked that in the first category, 88 percent of students were satisfied with regard of curriculum, students' participation and contextual test which gets the students in alive, motivated, but not isolated and choked to death.

We just need to remember that this category represents one aspect of the learning process and it's counted for one quarter of the evaluation. A total of 88 percent mentioned here would be all right because it is superior to 80 percent and it would be B+ as a letter grade. Unfortunately the last two other categories are below 70, the average level.

We have made this analysis based on the students' perspective, and we have found that the educational system would fail to instruct the students satisfactorily and to lead them to success indicating a state of educational catastrophe. It is now critical to conduct analysis of the teachers' data to see if there would be, at a certain degree, some differences in order to make a general conclusion.

Analysis of the Teachers' Data

Teachers' questionnaire is queasily similar to that of the students, and the reason why it is that is simply there is one single activity going on in one classroom at a time. That single activity is instruction and learning combined; teachers are giving instruction and students are receiving it. Because one person is giving it, we call it teaching and another person is receiving it, we call it learning. Therefore there will always be curriculum and its use to be comprehended, TM and its use to be also comprehended, and participation of the students which should be the teachers' teaching domains. This is in this platform that one would need to understand interaction between both teachers and students.

Based on relation that exists between curriculum and teaching and comprehension of these activities and types of assessments, it had been necessary that we asked the teachers questions about them to identify the kind of instruction kids were receiving as we have already checked the level of their comprehension for the good of the system.

It is fortunate that 91 percent of TPA confirmed that the state curriculum has been circulated around their classroom and communication-curriculum based in their school system has been perfect. Moreover, kids have been given time to discuss about their needs harmoniously. At that level the instructional system is champion for allowing it to happen. It means that the students have received adequate instruction to achieve success. Now, one of the greatest questions an analyst should ask is about reception of that exceptional instruction. Another great question would be about relationship between, again, that exceptional instruction and the quality exam given to the kids, including content and structure without forgetting the notion of method utilized by the teachers to facilitate comprehension of the students.

Again, 91 percent of teachers put emphasis on the importance of the use of a teaching method to reach the students' capacity of comprehending their teaching content and to check for their understanding. Most of them use teaching methods such as WSGI, VA, GR, and CCI (individual). It is good and excellent, but, we have not heard about any of the famous and successful methods that we have mentioned in the above paragraphs such as mastery learning of Benjamin Bloom and mastery teaching by Madeline Hunter which would be a plus for the educational system. Unfortunately, we are not sure if the in-use methods are up-to-date which is to be searched and investigated by us, or some other researchers.

Again and again, 91 percent of teachers affirmed that they always check for their students' understanding, have always provided feedback to them, and have been testing them on a regular basis as what they call formative assessment. They contended that there is no such serious instruction and real teaching activities out of those being mentioned. We think that teachers have revealed themselves as champions for articulating the highest level of instruction. However, not all of them think this way and cultivate this potential aspect of teaching. From 100 percent, 91 percent represent a high level of qualification indicated above; what about the rest: 9 percent left which might be contagious? This is what we need to think about. Why can't the system register full potential reaching 99 to 100 percent of perfection? Remember that even the teachers are so positive regarding their highest level of teaching and kids' participation and their method applied to teaching, most of the students are still facing failures. What exactly happens?

The answer to that important question may be found in the next and last category of answers to the group of questions asked in the teachers' questionnaire investigating relationship that exists between curriculum, class content, discussions, nature of the assessments. It seems that there is no match among them at all. Moreover, teachers are clear on types of questions that should come first and last. In addition, they have no notion of psychology of mind during exam times, and they think that it is good to trick the students disregarding possibility for them to fail and that students' failure is theirs.

It is reported that 10 percent of participants left questions with no answer which is a sign of ignorance. It means that they also ignore the notion of test structure, and that they are taking a chance when it comes to building an assessment. It seems that a real educator would have been necessarily a psychologist. In addition, it would be better for educators to have some psychology and psychometrics classes. A real educator should be one who puts himself or herself in kids' shoes in order not to cause their woes.

Another 10 percent of participants affirmed that all their exam questions are tough. This category also is acting as ignorant when it comes to structuring their assessment, and they need to pay attention to advice given to peers above. Education is not a game, and teaching is not a chance that one is taking; it is a sacerdotal career that one should warmly embrace and consider themselves as champs in athlete fields. When a teacher is leaving home to meet his or her kids, they should leave with winning determination and think that they can save the whole world by only helping a kid to be successful in school.

One thing we know that is certain, no student can be successful if the teacher does not shape structurally his or her assessment and if he or she does not test the students based on the class content regardless of anything keenly done in the classroom. The key is a package: teaching in the context of a state given standard curricula, discussion of that content, teaching with method, and structuring the assessments. There is no salute or success out of it.

Let's go back to the rest of the answers. The rest of the participants (80 percent) happily affirmed that they always start their test with easy or memory questions. They continued to say that, unfortunately, their students hardly earned a B; they remain C students or average ones. That is a poor statement. If more than 80 percent of the class fails to pass the exam with at least a B, the class fails the exam globally having in head that C is not an excellent grade whether it was considered as a passing grade in the actual system. Then, the teacher fails to be a champ. The grade of majority of the students is that of the teacher.

Conclusion of Data Analysis

Based on the analysis of the data, the great majority of students are satisfied with their teachers' instruction for one part. For the other part, teachers think that they are doing good job using the imposed curriculum, having the students' participation and discussing their class content and their method. However, findings prove that the students still can't pass the final exams, which remains the educators' greatest challenge that should be addressed with different instructional strategies, different teaching method, and different psychological approaches related to testing and the like.

It is found that there is lack of structure in tests given to the students. A great number of teachers agree that it is important to start the assessments with memory questions, which they (80 percent) are actually doing, however, they student still cannot earn a B grade, and this has been posed as the greatest problem and challenge of the educational system. The worst is that findings prove that some teachers, those we had a chance to talk with while answering the survey questions, intended to ask only tough questions and contended that it is even possible to trick the students by asking tricky questions to check out the degree of their intelligence, which is recognized as a traditional myth.

In this situation, the result has turned to be negative and catastrophic. No teacher had an idea of the structure of test design consisting in starting asking memory questions or easy ones and ending the assessment with memory once again while the middle of it should contain tough questions. Therefore, our hypothesis which stipulated or assumed that it is difficult for students to pass successfully an exam when that exam is not structured is justified and accepted to be true.

Recommendations

It is recommended that test or assessment designers utilize the bell shaped testing system (BSTS) proposed in the beginning of the research. It is a testing method based on which tests will be really made comprehensive with memory questions in the beginning, skills questions in the middle followed by progressive tougher and toughest ones. After different aspects of skills questions, based on Bloom taxonomy moving from simple to complex skill questions, have been displayed orderly, the test designer may add memory ones still based on Bloom's ideas of progressive elements to build a complete comprehensive test.

It is also recommended that test designers at state and particular instructors levels design their tests strictly based on the actual or taught curriculum content, or based on the part of the curriculum taught in such that students do not have to spend time on questions that were not viewed at a certain degree. They will never be able to provide satisfactory answers to those questions, except if they have read a lot, and if they have luckily made educated guess. Why should a teacher do that to his or her dear students?

In addition, it is determined that emphasis should be put on parts of teaching content where students have given their input. In this case, question should be built based on a class discussion to expel a kind of deep intelligence, and this question would have been part of skill or tough questions still out of trick. It is true that will appear to be kind of easy questions if they remember that it is made from their own entry. This is why a qualified teacher should pay attention to their students speaking, questions seeking understanding. When it comes to students, there is no stupid comments, intervention or participation, and or damn question. Teaching is symmetric or two-way communication, dialogue, in the Socratic sense, between a teacher and a student. Class discussion reveals itself to be the greatest context for test questions.

Lastly, there should be a selected contemporary working method of teaching or an innovative one. A teacher should not take any chance embracing a teaching career without having in advance a teaching method. Otherwise, that teacher is taking a chance of crossing the creek with no paddle, and that may turn to be a shame. Teaching is the act of building tomorrow's society, which should not be built anyhow negligently; teaching is a sacerdotal activity that the priest should embrace with heart and spirit. Always depart with a strengthening teaching method to lead the kids to heaven. May the activities be a two-way movement including not only a teaching method, but also a method of testing coping with the bell shaped testing! If teaching is the act that intends to show the kids the heavenly way and testing is to locate their position on pathway, checking for understanding might be the critical location badly needed (Hunter, 1994; Fisher and Frey, 2007; Wiggins and McTighe, 2011).

Do you wish not to lead the kids to any mire or to their cave? Then pay attention to those advices aiming to make you wise moving you away from the cusp of fall. Otherwise, you will fall in losing your balance, and you may not be able to stand anymore to nourish your greatest expectation and caress a paramount attempt sustaining teaching success with colorful fragnance. Then, implement those advices while listening to your heart.

CHAPTER IX

Evaluation of the Result

Evaluation implies profound epistemological reflections on the findings or results of a research topic in order to find criteria that allow the researcher to efficiently judge the research findings in relation to the hypothesis. It means that it is possible to check for the following criterial elements: validity, reliability, and accuracy of what happens to be a synthesis of all encountered viewpoints in the research that tend to contradict each other or to be a congruent point that one can stick with. We should look for generalizability of the findings because it is required in qualitative method, which mostly deals with particularity, according to Greene and Caracelli, introduced by Creswell (2009). Therefore, only validity, reliability, and accuracy will be used as established evaluative criteria of this research.

Remember that this step was announced in the first parts of the research, where it said that our main research method was *NAPOQMER*, a six-step research method whose objective is to help get deeply into a subject matter that is of interest to the population. "*E*" being the fifth step of that research method stands for "evaluation." It means that *R* is going to be the end of the process, when we will need to write the final report (*FR*).

Validity of the Research Findings

According to Creswell (2009), validity implies three main elements that are designed to help check if the data collected from the participants are credible. Those three elements are trustworthiness, authenticity, and credibility, which should be viewed separately.

a. **Trustworthiness:** Trustworthiness reflects the degree to which one can trust the document or what has been said. In the context of analysis of data, it is good to see if the source is trustful. Its trustworthiness indicates that the researcher has collected good information regarding the research topic. The information can be qualified as truth depending on the qualification of the source and if the person owning it or representing it is the right person to give that information. We just would like to inform that we have only spoken with the teachers as instructional dispensators; we have also talked to an elementary principal who embraced and supported the project with all his heart. He promised that he would have his teachers fill out the surveys. After more than a week, we received a big envelope containing the completed surveys by hand. This is the reason why we think, 100 percent, that those data are trustworthy and that we can use them.

Regarding the rest of the data collected, we met with the participants, whether teachers or students, face-to-face, and we had the opportunity to have them fill the surveys. Some of them talked to us while filling them out. We also tried to talk to them in order to make sure that they knew what it was about. The greatest remark that one could make is that the participants revealed to be authentic. It is again the reason why we place our confidence in the data collected and in the sources. What about their authenticity?

b. **Authenticity of the Data Collected:** Lagemann and Shulman (1999) contended that the data provided by the informant (participant) represent his or her view of the world and interpretation of the events. If the informant communicates everything he or she knows it is because he or she trusts him or her and never thinks about betrayal. In other words, the informants identify a similarity between their view of the world and that of the researcher, as it could also be identical when it comes to interpretation of the data. Therefore, they have no reason to withhold the truth.

However, Lagemann and Shulman (1999) still thought about probability for the informants to provide wrong information to the researcher because of lack of trust. This is the reason why they distinguished between authentic and false findings. They wrote the following: "By authentic findings, I mean that the data represent an informant's actual way of seeing and interpreting events; they represent truth at least as far as the informant understands and is able to communicate his or her truth" (P. 230). Out of this trustful relation which is based on what the researcher is, what he or she represents in the community where the investigation takes place, and the way the researcher presents him/herself, the informants would not provide any trustful or any information at all (Lagemann and Shulman, 1999). In addition, it sometimes depends on the topic subject and its importance for the community. In this case, the researcher had full access to his informants' data or world not only because of trust but also because of the topic subject, which is a reflection on the method of testing students who are victims of a failing educational system. Therefore, the data provided by our informants are authentic.

The *Newbury House Dictionary of American English* defines authenticity as "the genuineness and validity of something." It means that someone who lets other people get into his or her inner world also lets them get the wholeness of whatever he or she knows as truth. This idea leads us to what Northouse (2007) called integrity, which he defined as "the quality of honesty and trustworthiness" (p. 20). To come back to the subject— reflection on trustworthiness and validity of the data collected—we can say that we saw sincerity and integrity in the informants' faces as they were eager to fill out the survey and let us access their inner rooms. Therefore, the data collected from the informants are worth reporting.

Authenticity is, lastly, a state of what is real, having no fake or fading color. When one tries to state what exists and is true, other people or the listeners can see and identify it quickly with no special effort. It is in this perspective that Peter Block, in his consulting activities, reported his authentic statement and declared that his clients were very satisfied because they felt that Block (2001) was telling them about their lives and expressed willingness to tell the truth. This is what authenticity is all about. Here's what Block (2001) wrote: "There is no idea that has been more useful in my counseling experience than 'authentic statement,' putting what one is experiencing into words at the moment they are being experienced" (p. 253). We have experienced this fact as much as our informants; they felt that we have met them halfway, and we've been found not to be many anymore, but as one and a single soul. Here's what the gatekeeper of the elementary school told us: "I can't wait to read that research." This is where we find trustworthiness and authenticity of the data collected. What about their accuracy?

c. **Accuracy of the Data**: We admit that the participants could have made mistakes, but not willfully because we have stated that they were willing to give us information based on their ability and what was within their reach. One thing that appears certain is that the subject was not something from the sky but a day-to-day subject: education focused on the testing system and search for the cause of failure of the system.

One problem might be related to our sampling size. It is really small having only 50 participants in Florida. Fortunately, some other researchers may use a larger sample size to investigate the same subject, which is of great interest to the entire population. However, to our knowledge, our participants were truthful and authentic; it is very possible that they were also accurate. Who knows?

Implications

Bell-shape testing system (BSTS) will undoubtedly be a part of the educational system in all respects and influence all stakeholders when it comes to making decisions related to the programs, curricula, teaching methods, and test designs. The research findings will impact the educational system in Florida, in the United States, and in the world where every educator—including board members, test designers, principals, and teachers—will be able to consider this testing system at the moment of their decision-making processes-assessment related. It will be the salutary path of the failing system that lasts so long, which deserves a chance to be changed and transformed such that educational outcomes could be better. It will be an urgency upon reading of the research paper because it is not wise to keep playing with a losing team. Therefore and fortunately, educators will find a way not to belch at night before they sleep without reviewing what has happened during the day that is not favoring the system for the spirit of structure will not leave them at rest (Collins, 2001).

This text should be considered as a wake-up call. Educators have been sleeping too much; they keep sleeping while it is already daylight. The system is very late, but it is not too late. Therefore, hindrance related to the process of change is not advised. And whatever we need to do shouldn't be postponed at all for tomorrow. Tomorrow will take care of itself. Let's do whatever we have to do today. Let's drink some instant coffee so we do not have to procrastinate; we are urged to act now and only now.

When considering and implementing this innovative way of teaching and testing, it may appear to be a revolution in the educational system, a global phenomenon at all instances—including elementary schools, high schools, technical schools, colleges, and universities. The view should be that broad because education and instruction take place everywhere there are human beings, societies where transformations are taking place or want to take place. Education implies transformation. If education has taken place and no transformation has happened, then education was pseudo-education. Real education has this goal: transforming people and moving a society from mediocre to perfect, which could be highly effective by implementing the BSTS.

It is unfortunate to read the responses of some students, those who say that they never received feedback from their teachers and they don't know the meaning of feedback. Those students are lost in the system and will never understand their teachers because understanding of new notions depends on old knowledge received in previous lessons. Unfortunately again, some of them would never ask questions when they miss an explanation. Therefore, it is up to the teacher to check for understanding. In this, a lot of teachers are on the verge of failure.

Some teachers affirmed that they teach based on the curriculum and use a teaching method, check for their students' understanding, and provide feedback to them. However, they still can't earn 80 percent; some students still get Cs. It is unfortunate and inacceptable. All kids can learn and make it happen at the same level. In order for this to happen, they need to be taught in the same environment, have the same learning opportunities (Carroll, 1963; Bloom, 1971; and Engelmann, 1985), and be tested using BSTS as the reinforcement of well-applicability of fairness and educational technicality.

It is advised that teachers need to be greatly concerned about success of the students, and be intrinsically motivated to teach at the highest level possible using great teaching and testing methods while considering BSTS as one their greatest alternative when it comes to technicality. Anyway, they need to be charged with responsibilities of the kids' failures as long as they have received adequate training and necessary tools to teach based on a standardized manner. Principals and assistant principals may address it all their career as militants and practitioners. However, the teachers also need to be integrated in the administration's decision-making system. This is now only that improvement possibility may appear to be effective and fully certain.

Those are genuine administrative steps to help get the situation in hand, but what are the technical tools teachers need to master? They need to understand that they will not make any progress out of the state curriculum, out of the utilization of a winning teaching method that could be the most working one (such as the mastery learning of Bloom, the mastery teaching of Hunter, or that of Engelmann, even though it is a little bit complicated, with a proven winning teaching method. These methods put emphasis on the teaching content—curriculum related, students' understanding, feedback, formative assessment.

Furthermore, teachers could be trained on quality test design, which refers to the bell-shaped testing system as exposed and explained above, which is going to be the master key of teachers' and students' successes. It contends that no student can be successful if they are not strictly taught based on the curriculum content, based on their input in the decision making system (participation of the students at the moment of the course plan presentation) related to the content. Not only those, they cannot be successful if they are not assessed based on the current teaching. If teachers pay no attention to the prescribed method of test design (BSTS). We urge them to heed it.

One other thing is that the result of this research will be able to fight the capacity of the failing system to hinder the process of making all students successful because of its generation of lack of democratization of education. It was remarked, as a fact, that not all kids have received the same privileges. Some of them were set for failures, and some others for success (English, 2008). Therefore, this research's results will somehow motivate educators to understand that there is a need in this neglected area to democratize education and teach all the kids equally likely.

Educators and teachers will recognize that it is time to individualize education and to heed to the minorities. Those kids have equal potential or intelligence with those who grew up in exceptional environments. Therefore intelligence belongs to all kids, but not some of them; they just need attention and affection, special touches related to their low background's environment (Gardner, 1995). Findings prove that kids who receive attention and affection experience greater neuron development and increased neocortex to become more intelligent than ever (Panksepp, 1998). This is to start thinking about the plasticity of the brain, the siege of the real learning process being the subject of our next research. One way of showing love, affection, and attention to kids is teach over and over the part of the course they do not understand.

Regarding this, Benjamin Bloom contends that if a kid did not understand a class presentation, he or she needs to be taught again and again until he or she reaches full understanding, which is a necessity. These are all it is about. We, educators, principals, teachers, and test designers, need to think about it now and forever as long as we want to build a normal and great society. Our kids today are called to be, with no exception, our tomorrow's senators, representatives at the lower chamber, secretaries of state, governors, and presidents.

Carroll (1963), Bloom (1971), and Engelmann (1985) believe that all our kids can pass their test at 80 percent or more if they are taught properly, and learning opportunities are given to them if they are not set for failure. Intelligence is well distributed at birth as birthright (Gardner 1995). It means that a superior IQ (intelligence quotient) is a kind of myth; higher IQ is related directly to the environment in which kids grow up and the learning opportunity they have been given, small or big. Some kids don't have any. Otherwise, some kids or students may experience what Howard Gardner called a kind of neurobiological and developmental constraints, which make them become rote and intuitive learners. This problem should be addressed as soon as possible.

A Narrative Report of the Findings

This research was an attempt to understand the U.S. educational system and to make a claim regarding its failure to properly educate and assess the kids. Therefore, it was also an attempt to present a salutary alternative aiming profound changes and an effort to save that failing system. We should not be considered as the sole accusatory researcher because Benjamin Bloom, Howard Gardner, and David Krathwohl in his *Taxonomy of Educational Objectives, Handbook II: Affective Domain* had, before us, pointed their fingers on it. The *toile de fond* relates especially to the testing system. Because there is no testing without teaching and that the test questions would be based on the teaching content. Educational Services and Research Center has tried to put them altogether and test the system. Unfortunately, it was found that our educators, teachers, and educational state representatives fail to lead the students to success.

Our hypothesis was the following: H1: If the teachers teach based on the established curriculum and use a good instructional method, students who are tested only based on what they have learned from their teachers and also based on bell-shaped testing—where questions are arranged from the bottom to the top, from simple to complex questions, or from retrieval to synthesis questions—will always earn grades no lower than 80. The hypothesis has been proven right and accepted as a fundamental truth. Therefore, the null hypothesis was disregarded as an idea with no foundation.

In fact, findings prove that teachers did not use proven and successful instructional methods and some of them neither checked for students' understanding nor provided feedback to them. In addition, teachers did not structure their tests and used mostly, or only, skillful questions. Therefore, it would be extremely difficult for the students to be successful on the tests.

Closing Thoughts

This result comes to, unfortunately, support previous research findings that have already alarmed failure of the education's system (English, 2010; Gardner, 1996; Wiggins, Reeves, 2007). Wishes would be that things could be better and a little bit different. Unfortunately, they are what they are. There is alarming necessity for change because one cannot keep kicking it with something that is not working. The bell-shaped testing system was built upon a cohort of teaching methods presented by some famous theorists whose names sound as manifests and are recognized as the greatest theorists of the century if not all times. Is it a way of honoring Madeline Hunter and her brother Robin Hunter, John B. Carroll and his editor and presenter Lorin W. Anderson, Siegfried Engelmann, Robert J. Marzano, Timothy Walters, Brian McNulty, Erick Erickson, Thomas R. Guskey, John Dewey, David W Johnson, Roger Johnson, Edythe Holubec, Paul Black et al., Robert M. Gagne, Karl Popper, Ellen Lagemann and Lee Shulman, put apart, L. Vygotsky, etc.? Their contributions have been huge, and this research was possible because of them.

As we have utilized a small sampling size (30 students and 20 teachers) and triangulation has not been utilized as an extra research to gather more data and verify the findings, it is recommended that more researches be conducted in this area. However, replication may not be possible because we have applied a qualitative research method (Creswell, 2009).

Acene Fleurmons, BSW, MOM, and EdD
Founder and President of Educational Services
and Research Center

BIBLIOGRAPHY

Argyris, C. (1994). On organizational learning. USA: Blackwell business.

Black, P. et al, (2003). Assessment for learning: Putting it into practice. Philadelphia: Open University Press.

Biesta, G. J. and Burbules, N. C. (2003). Pragmatism and education research: Philosophy, Theory, and educational research series. USA: Rowman and Littlefield Publisher.

Block, P. (2000). Flawless consulting: A guide to getting your expertise used (2nd Ed). USA: Pfeiffer.

Bloom, B. S. (1956). Taxonomy of educational objective: Cognitive domain. Chicago: Longman.

Bloom, B., Hastings, J. T., and Madaus, G. H. (1971). Handbook on formative and summative evaluation of student learning. USA: McGraw-Hill.

Brookhart, M. S. (2010). How to assess higher-order thinking skills in your classroom. USA: ASCD.

Bruner, J. & Anglin, J. M. (1: W. W. 973). Beyond the information given: Studies in the psychology of Knowing. N. Y: Norton and Company.

Buchler, J. (1955). Philosophical writings of Peirce. USA: Dover Publications.

Burke, K. (2010). Balanced assessment: From formative to summative. USA. Solution Tree.

Carroll, B. & Anderson, L. W. (1985). Perspectives on school learning: Selected writings of John B. Carroll. Hillsdale, NJ: Lawrence Erlbaum Associates.

Collins, J. C. and Porras, J. I. (2001). ....

Comenius, J. A. (Y. Not Revealed). Youth demands good education and right instruction- Pamphlet. USA: Kessinger Publisher.

Cone, J. D. and Foster, S. L. (2006). Dissertation and theses from start to finish: Psychology and Related field (2nd). USA: American Psychology Association.

Descartes, R. (1637). Discourse on the method. France: Larousse Classics.

Dewey, J. (1944). Democracy and education: An introduction to the philosophy of education. United States of America: Free Press.

Dewey, J. and Jackson, P. W. (1990). The school and society-The child and the curriculum. Chicago: The University of Chicago.

Dewey, J. (2012). Moral principles in education. USA: Publisher not revealed.

Elliot, S. N. et al. (2012). Handbook of accessible achievement test for all students: bridging the Gaps between research, practice, and policy. USA: Springer.

Engelmann, S. and Carnine, D. (1986). Theory of instruction: Principles and applications. USA: ADI Press.

Erickson, E. H. (1968). Identity youth and crisis. New York: W. W. Norton & Company.

Erickson, H. L. & Tomlinson, C. A. (2002). Concept-based curriculum and instruction for the teaching classroom. USA: Norton Company.

English, F. W. (2010). Deciding what to teach & test: developing, aligning, and leading the Curriculum (3rd Ed). USA: Corwin.

Fisher, D. & Frey, N. (2007). Checking for understanding: Formative assessment techniques For your classroom. USA: Association for Supervision and Curriculum Development.

Gagne, R. M. (1988). Essential of learning for instruction (2nd Ed). Englewood Cliffs, NJ: Prentice Hall.

Gardner, H. (1995). The unschooled mind: How children think & how schools should Teach. USA: Basic Books.

Gardner, H. (1999). Intelligence reframed: Multiple intelligences for the 21st century USA: Basic Books.

Gardner, H. (2006). Five mind for the future. USA: Harvard Business School Press.

Glaser, B. G. & Strauss, A. L. (1967). The discovery of grounded theory: Strategies for qualitative research. San Francisco, California: Observations.

Guskey, T. R. & Bloom, B. (1997). Implementing mastery learning. Kentucky: Wadsworth Publishing Company.

Heifetz, R. (1994). Leadership without easy answer. U. S. A: Harvard University Press.

Hobbes, T. and Propiel, J. J. (1651/2004). Leviathan. USA: Barnes & Noble.

Hoyle, J. R. et al. (2005). The superintendent as CEO. USA: Corwin Press.

Hunter, R. (2004). Madeline Hunter's mastery learning teaching: Increasing instructional Effectiveness in elementary and secondary schools.

Hunter, M. (1994). Enhancing teaching. USA: Mcmillan College Publishing.

James, W. (1950). The principles of psychology. Vol. 1. USA: Dover Publications.

James, W. (1991). Pragmatism. USA: Prometheus Books.

Jensen, A. R. (1973). Educational differences. Greta Britain: Constable Ltd.

Johnson, D. W., Johnson, R. T., and Holubec, E. J. (1994). Cooperative learning in the Classroom. Alexandria, VA: Association for Supervision and Curriculum development.

Juron, J. M. (1992). Juron on quality by design: The new steps for planning quality into goods And services. USA: Maxwell Mcmillan.

Kuhn, T. S. and Hacking, I. (2012). The structure of scientific revolutions (5th Ed). Chicago: University of Chicago.

Lagemann, E. C. & Shulman, L. S. (1999). Issues in education research: Problems and Possibilities. San Francisco, California: Jossey-Bass.

Lunenberg, F. C. and Ornstein, A. C. (2004). Education and administration: Concepts and Practices. USA: Thomson Wadsworth.

Maslow, A. H. and Bennis, W. (1998). Maslow on management. USA: John Wiley & Son.

Marzano, R. J. (2001). Designing a new taxonomy of educational objectives. California: Corwin Press.

Montesquieu, C. (1748). The spirit of the laws. France: Digitreads.com.

Molly, L. C. (1990). Vygotsky and education: Instructional implications and applications of Socio-historical psychology. USA: Cambridge University Press.

Northouse, P. G. (2007). Leadership: Theory and practice (4th Ed). USA: Sage Publication.

Perlstein, L. (2007). Tested: One American school struggles to make the grade. New York Henry Holt and Company.

Popham, W. J. (2003). Test better, teach better: The instructional role of assessment. USA: ASCD.

Popham, W. J. (2008). Transformative assessment. USA: ASCD.

Popper, K. (2002). The logic of scientific discovery. London and New York: Routledge.

Reeves, D. et al. (2007). Ahead of the curve: The power of assessment to transform teaching and learning. United States of America: Solution Tree.

Reeves, D. B. (2004). Accountability for learning: How teachers and school leaders can take Change. USA: Association for Supervision and Curriculum Development.

Russell, P. and Ferguson, M. (1979). The brain book: An engrossing application of brain research To self-improvement. USA: A Plum Book.

Sacks, P. (1999). Standardized minds: The high price of America's testing culture and what we Can do to change it. Cambridge, MA: Perseus Publishing.

Sartre, J. P. (1943). Being and nothingness. France: Gallimard.

Skinner, B. F. (1974). About behaviorism. USA: Vintage Book Edition.

Taba, H. (1962). Curriculum development: Theory and practice. USA: Harcourt, Brace and World.

Tileston, D. W. (2004). What every teacher should know about: Learning, memory, and the Brain. USA: Corwin Press.

Touraine, A. (1995). Critique of modernity. Cambridge, MA: Blackwell.

Vygotsky, L. S. et al (1978). Mind in society: The development of higher psychological processes. USA: Harvard University Press.

Wiggins, G. and McTighe, J. (2011). The understanding by design: Guide to creating high quality Unit. USA: ASCD.

INDEX

A

abilities, 29, 30, 32, 33, 41, 44–45, 55, 78, 79, 110
abstraction, xviii, 31, 73
adequate yearly progress (AYP). *See* AYP (adequate yearly progress)
adjustment, 7, 8
affection, 112
analysis
  light, 41
  profound, 41, 97, 99
application, 29, 32–33, 35–36, 38, 44–45, 47–48, 75–76
assessments, xxi, 1, 3, 5–6, 8–9, 11–12, 15, 17, 19, 21, 23, 25, 27–30, 35, 37, 49–51, 58, 60, 72, 76, 88, 92–93, 95, 97, 102–5
  common, 24
  formative, x, xx, 1, 2, 3–12, 14–15, 19–22, 24, 26, 42, 52–53, 61, 63–64, 80, 83–84, 92, 98, 103, 112
  higher-order thinking, 5
  interim, 9
  needs, xxi, 3, 16
  short-cycle, 9

summative, v, xx, 1–3, 5–6, 8–12, 14–15, 18, 20, 22, 24, 26–27, 30, 42, 64–65, 83–84, 95
assessment standards, 18–19
assessment structure, 27
assessment system, 17, 20
assessors, 6, 13–14, 18–19, 28, 37, 46, 48, 49–50, 81
authenticity, 107–9
authorities, 67
automaticity, 20
automatic memory. *See* memory, automatic
AYP (adequate yearly progress), 9, 22–23

B

behaviors, xv, 30–31
beliefs, xiv, 37, 87
bell-shaped structure, 84
bell-shaped testing (BST). *See* BST (bell-shaped testing)
bell-shaped testing system (BSTS). *See* BSTS (bell-shaped testing system)
birthright, xvii–xviii, 55–56, 113